THE BRIGHTON GARRISON

1793 1900

Published by R C Grant

ISBN 0 9530332 0 1

Printed by CPO Print, Worthing

CONTENTS

ILLUSTRATIONS

Drill Halls and Local Volunteers

Preston Cavalry Barracks

Preston Barracks Hospital

ACKNOWLEDGEMENTS

I am most grateful to a large number of people and concerns for their help in putting this book together. Apart from my family who had an awful lot to put up with because of my frequent absences, enthusiastic ramblings and squeals for help as the computer seemed to irretrievably gobble up a day's input, numerous public institutions and private individuals all played a considerable part by providing material for its compilation. I was always welcomed by the staff in the reference sections of the libraries in Brighton, Hove, Worthing, Chichester, Sussex University and the British Library in Colindale. Further thanks go to those in the local and National Public Record Offices in Lewes, Chichester and Kew, and Dr Boyden of the National Army Museum in Chelsea. John Roles and Brighton Museum's Art department provided tremendous assistance by giving access to resources not normally available and local military enthusiasts also contributed by offering more specific expert guidance. The human touch was added by my many colleagues within the vast membership of the Sussex Family History Group, who not only provided stories but allowed treasured photographs of their military ancestors to be published. Mr Peter Goddard at the printers also helped by patiently guiding me through the publishing maze and handling all those small obstacles and large illustrations which I found myself totally unable to cope with. Finally, although he only contributed one short paragraph, a special thank you goes to Mr. Spike Milligan, who just rang me up and chatted in person about his own experiences in the Victorian cells of Preston Barracks, and by doing so, made my day.

INTRODUCTION

Why I set about this task

Finding out that Preston Barracks were being demolished came as a total shock to me and I instinctively rushed out with my camera and shot off the last few frames of the only film I had. Only in retrospect, did I realise that my initial and rather amateurish token attempt to record a piece of important local history was grossly inadequate and looking back now, I regretted that I had squandered the many opportunities that had once been open to me. After all, I had spent a considerable time inside Preston Barracks during the six years I was in the Territorial Army in all that time I never once considered making a permanent record of the many unique buildings with distinctive nineteenth century military architecture. Jobs like that were left to *"someone else."*

Fortunately, when the first phase of demolition was over, the contractors left the end two barrack blocks standing and like Timothy Carder, the author of the Encyclopaedia of Brighton, I presumed they would remain. Lulled into a false sense of semi-contentment, I believed that these buildings at least would provide an ideal opportunity for that *"someone else"* to make a detailed record in the future. The matter was then put aside until regrettably, it slipped from my mind completely.

Sadly, I had not realised that the reprieve for these two historically important buildings was only temporary and that they too would soon surrender to a powerful heavy ball on a chain and a twentieth century bulldozer. Bitterly disappointed, I turned to the local archives for information, only to find that with the exception of the a handful of illustrations and supporting text on the town's coastal battery, the local reference library held little in the way of documents which recorded Brighton's early Military history. Virtually nothing was on file which referred in any depth to Preston's Cavalry Barracks and it seemed that not one military photograph had ever been catalogued. When I widened my search, I made the further discovery that the former Infantry Barracks which once stood off Church Street, had also suffered a similar fate and that too had been demolished without a detailed record being made. Even further disappointment came when the staff at the Borough Surveyor's office advised me that they did not hold plans of existing or former War Office properties or other structures crucial to the supply of essential services, for access to much of that information was denied to them under the "Crown Properties Act." It soon became apparent that anyone with the slightest interest in Brighton's military past would have to conduct a particularly wide search and seek out information which over the years had become scattered in the more diverse texts of numerous personal diaries, or was buried deep in the Public Record office beneath countless administrative papers for the South Eastern Command. To me, this represented a serious gap in the local archives and I had this feeling that someone ought to produce a accessible account of this important facet of Brighton's past. There was a need for a simple reference document which included a résumé of some of the available writings, plans and illustrations, and anything which would give researchers and historical enthusiasts a point from which to gain greater access to a more extensive range of local information. Not really being one to motivate others and certainly no expert in military matters, I nevertheless set about the task myself, responding to my own challenge and collecting together as many pieces of this gigantic jigsaw as I could. I hoped at least to reconstruct an outline of Brighton's former military history. Realising it was to be the first document of this nature, I felt that if there were areas where any of my views, opinions and interpretation of the evidence were considered inaccurate, it would promote constructive discussion and that too would effectively further augment the research.

I had not realised how rewarding the task could be for it began to provide more details about local military buildings and introduced me to former military sites that until then I had not known existed. My investigations also began to unearth highly fascinating stories about local military characters and events which until then had been unknown to me. As my collection of nineteenth century plans and illustrations grew, the formerly rural landscape around Brighton was revealed, captured in drawings showing the turmoil of mock battles. Furthermore, early writings disclosed that Brighton was of far more importance militarily and featured in art, music and literature. Of special interest were several sketch plans of Church Street and Preston Cavalry Barracks which were once considered as classified information. These were bundled in stacks of dusty government papers which in my opinion suggested that local Victorian officialdom welcomed the officers, but regretted the presence of the common soldier in the town. Perhaps that is why I eventually settled on the use of the word "Garrison" in the title for this book, because the term was sufficiently flexible to encompass a study of people as well as places, even though almost all of the latter had now disappeared without trace. Rather than dwelling on what cannot be undone however, my intention here is to mainly to provide that initial access to references about Brighton's nineteenth century military history, not to question why so many unique military structures were initially neglected and when once beyond economic repair, were demolished without first being properly documented. Although the colourful regiments have long since departed and the many unique military buildings they left behind are now irretrievable, it is still worth remembering that throughout the nineteenth century the presence of the military in Brighton played a considerable part in the geographical and social development of the town.

Setting the Scene

Up until the sixteenth century, Brighton was just one of many insignificant little south eastern coastal villages with no major economic or strategic importance and the community were most vulnerable to attack from pirates and other invaders from across the Channel. In those days the town was not protected by any residential militia and in times of crisis, the community undertook the defence of their property themselves. Their instructions were to dig trenches and otherwise bar the streets with timber, furnishings and carts and then harry the enemy in any way they could. The British Museum has an early drawing, reputed to be either from 1515 or 1545, which shows the town being sacked by the French and then being put to the torch.

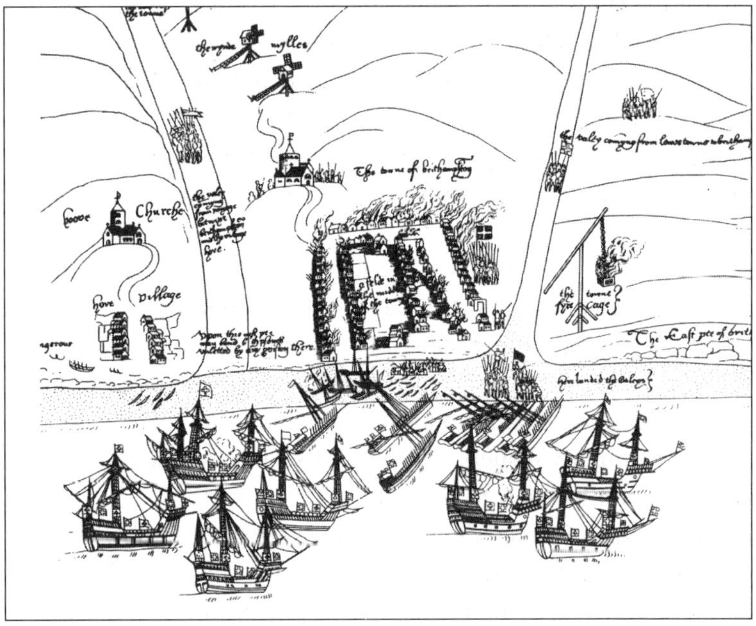

An Attack on Brighton by the French Fleet (1545)

The sketch, drawn on vellum, illustrates that the town was virtually defenceless as the population could only place their faith on the fire beacons which when lit, drew the farm labourers in from the surrounding area. These untrained civilians were then responsible for keeping the more organised enemy at bay until a troop of soldiers arrived from as far away as Shoreham or Lewes.

By 1559, Brighton had developed a substantial fishing fleet of well over 80 boats and this increased prosperity continued to heighten the local people's fear of their vulnerability to an attack. To ease their minds and protect his own investment in Brighton's future, the Lord of the Manor made over a narrow strip of the foreshore to the local fishermen who agreed to commit a quarter share of their profits on each catch to fund the building of a defending wall and a coastal battery. The resulting fortification of the village of Brighton's urban frontage was a 15ft high stone wall some 400ft long with a 3ft high parapet along its entire length. At the eastern end of this extensive wall was a gate-house tower and on the other end (opposite Middle Street) was a fortified battery which became known as "The Blockhouse." This coastal fortification was continually manned by the local fishermen who regularly volunteered to take turns to oversee the defence of the village. The 18ft high Blockhouse had 8ft thick walls and measured about 50ft in diameter. The 34ft circle inside the building was separated into several rooms in which chain-shot and powder were stored. Its roof was supported by huge stone arches which gave the building the additional strength needed by the gun platform above. The armament consisted of 4 large cannon which a local man, John Slutter was leasing from the Tower of London and numerous smaller pieces of ordnance (called Qualivers) which had been taken as prizes from foreign ships. By 1658 the building was being used as a prison and this continued until 1705, by which time, much of it had become unsafe because of the sea's constant erosion of Brighton's western coastal cliff. In later years the defending wall and parapet were lost to the sea and by 1749, the Blockhouse and gun garden were also so badly undermined, the majority collapsed onto the foreshore. When the coast road was reinforced in 1761, little evidence remained of the original building or the substantial fortified wall which had previously formed the backbone of the town's defences.

The Remains of the Brighton Blockhouse in 1761

More detailed information on the Blockhouse and the Eastern Battery which replaced it appears in Frank Kitchen's extensive article published in Vol. 21 of May 1986 edition of Sussex History.

In 1759, work began at the southern end of Great East Street to build a new coastal battery which had been commissioned to replace the Blockhouse. The plans which were estimated to cost £995, required the town's East Gate to be demolished and a two story house for the Master Gunner to be built.

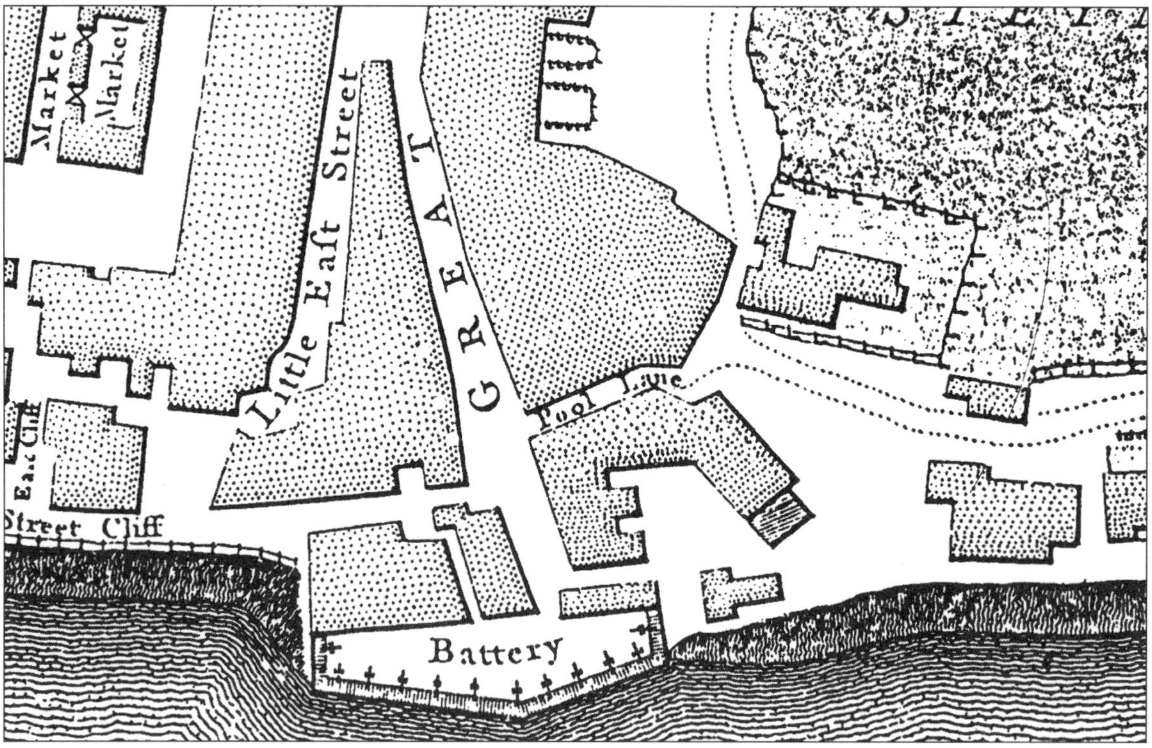

Plan of the Great East Street Battery: 1759-1786

The new Battery, which included a room beneath it for the storage of ammunition, was fitted out with 12 rather ancient cannon which were each expected to fire a twenty pound shot. Their effect was largely cosmetic for although they were reported as being satisfactory when inspected in 1779 by Thomas Bloomfield, who was said to be an artillery expert, the gunners themselves knew the guns were in a most dangerous condition. The wooden carriages were rotten in places and several of the guns were so riddled with rust that only four of them were considered to be safe. An unofficial recommendation was that they should only be used for Royal salutes. Even that proved to be over optimistic, for in September 1782, the master gunner lost both his hands whilst reloading during a salute to the Prince of Wales' great aunt Amelia, and early in the September of the following year, the under gunner was mortally wounded after a cannon ignited prematurely during a similar salute to commemorate another visit made by the Prince of Wales. The explosion blew the gunner's body some distance onto the beach, shattering one of his arms and removing his hand, which was never found. This battery also fell victim to encroachments by the sea and finally collapsed during a severe storm on 17th November 1786. The records show that 17 barrels of gunpowder were stored in the magazine at the time, but there was fortunately no explosion even though the roof and two heavy guns, each weighing over two tons, tumbled in.

With both the Blockhouse and its replacement Great East Street Battery destroyed, Brighton existed for the following seven years with no effective coastal defence or armament. The town and its fishermen were once again vulnerable to Dutch and French pirates who periodically attacked the local shipping. Furthermore, it was becoming generally noticeable that the families of the more astute members of the French aristocracy were beginning to arrive in the town and it was rightly believed that it would only be a matter of time before the citizens of the newly developing French Republic sought to expand and perhaps even overthrow other monarchies as well as their own. If that was to be the case, in the wider interests of national security, it became especially important to establish a permanent military presence in many key sites on the south coast. However, it was only after war was officially declared between England and France in February 1793, that action was finally taken to ensure that much of the southern coastline was better protected against a possible invasion by Napoleon. Funding became immediately available to update military facilities and work started on building further fortifications along the coast. Numerous canvas tents hastily erected on the hillsides outside towns and villages on the southern coast were quickly occupied by thousands of regular soldiers and men newly recruited into volunteer militias who were being drafted in to affirm England's military might to the potential aggressor from across the channel. During the more inclement winter months however, many of the soldiers had to leave their canvas accommodation and be billeted on licensed premises, an arrangement which was known from past experience to be highly unsatisfactory. It was therefore decided to upgrade the plans for coastal defence and include the construction of numerous temporary timber barracks in those towns thought to occupy positions essential to the country's defence strategy.

The continued speculation about Napoleon's army landing at Brighton resulted in a total review of the town's defences and a more permanent military presence being established for the duration of the hostilities. Two new town batteries were commissioned to replace the old gun gardens which had been eroded away by the sea and construction began on three temporary blocks of barracks which would provide winter quarters for a locally based infantry regiment. Just outside Brighton in the rural village of Preston, a further barrack complex was commissioned to provide more substantial accommodation for a full regiment of cavalry and supporting units of field artillery.

By the winter of 1793, Brighton had become a hive of military building activity with construction work taking place on;

[a] Two new coastal batteries to defend the town from an invasion force.
[b] The West Street Barracks and the East Street Cavalry Guard House.
[c] The Windsor Street Barracks and the Church Street Military Hospital.
[d] The Church Street Infantry Barracks.
[e] The new Cavalry Barracks on the Lewes Road in the parish of Preston.

Most of these structures failed to last for very long. West Street barracks and the guard house in East Street were de-commissioned around 1802, Windsor Street Barracks and the military hospital were sold off around 1815 and the sea claimed one battery whilst a road widening scheme claimed the other. Although Church Street Barracks was neglected and poorly maintained, it remained in use until 1869 when it was demolished, only leaving the more prestigious Cavalry Barracks in Preston to survive into the late twentieth century.

The village of Brighton had initially gained prominence as a seaside resort through the efforts of Dr. Russell, but because of the threat of an invasion, it also became a small garrison town. Credit must therefore go to Napoleon, for he indirectly gave the British soldier a first permanent foothold in the town and established a small but important military presence that was to continue for almost another 200 years.

EARLY ENCAMPMENTS 1793-1803

Following the declaration of war against the French Republic in February 1793, a number of temporary encampments were set up along the south coast which were occupied by regular soldiers and volunteers from the county militias. Primarily these camps were used to hold extensive manoeuvres and pre-orchestrated sham fights with the aim of boosting local morale. Brighton, formerly a small seaside village, was at that time without a barracks or a permanent military presence even though the Lord Lieutenant the Duke of Richmond fully realised that it occupied the shortest possible route from the English Channel to London and could be crucial to the strategy of a Napoleonic invasion force. It is suggested that this weakness in defence induced the Duke of Richmond to centre many of the military exercises in the Brighton area, for over the following years the fields on the village outskirts became enormous tented army camps accommodating regiments of Militia from all over the country. To assist in what today's strategists might imply was an overall illusion that Brighton had developed into a military stronghold, the city press sabre rattled and provided frequent coverage of the activities of the Prince of Wales and his own regimental guard from the 10th Light Dragoons which boosted national confidence. However, whilst the defensive value of each exercise was appreciated, the Prince himself and many of his more senior officers, flippantly treated the regularity of the Spring and Summer encampments in Brighton as an attractive addition to their social scene.

The First Brighton Encampment (1793)

The only known contemporary plan of the very first Brighton encampment (which was commonly known as The 1793 Downs Camp) is still preserved in Worthing Museum, but in a very peculiar format. It shows three maps of that year's series of encampments at Waterdown, Ashdown and Brighton & Hove, but most unusually, they are all printed on the fabric of an ordinary fan, which is thought to have been produced for sale as a souvenir for an officer's lady.[1] *Although the ravages of time have since obscured some of the details on the original fan, Andrew Goodwin's 1986 copy of the Brighton section of the three maps provides a far clearer illustration of the information on it.*

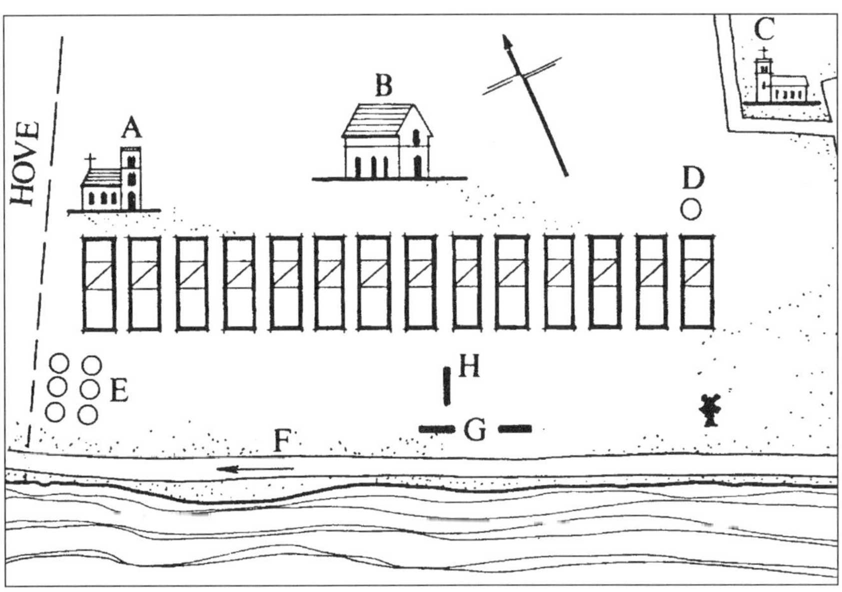

KEY
- A Wick Church, (St. Andrews east of Hove)
- B The Magazine
- C St. Nicholas — the old parish church of Brighton
- D Lime Kilns (at the bottom of Church Hill)
- E Tents and Pavilions of the Duke of York.
- F The Road to Shoreham
- G Lime Kilns
- H Ovens

The 1793 Downs Camp.

The regiments accommodated along this west to east strip were;

1 The Prince of Wales' 10th Royal Dragoons
2 The Light Artillery
3 The East Norfolks
4 The Berkshires
5 The West Essex
6 The Oxfords
7 The North Devons
8 Westminster/Middlesex
9 The West Suffolks
10 The South Devons
11 The Sussex
12 The North Hants
13 The Dorsets
14 The East Middlesex.[2]

The Volunteer Militia supported by artificers, regular troops and heavy artillery, were dispatched to exercise in Brighton on the 5th August 1793 by the Duke of Richmond and marched into town from their previous encampment at a training area in Ashdown Forest. The military influx was such that the fields around the town thronged with up to 10,000 troops and at a stroke, the local population was more than doubled. In addition to the astonishing number of volunteers and soldiers, many of the troops brought their wives and children with them and their presence attracted a host of merchants, petty tradesmen and numerous other camp followers. Although exact figures have been difficult to establish, it is generally estimated that when the encampment was at its peak, the population of Brighton had trebled. Any analysis of population statistics however, could never be accurate for it was even further complicated by the periodic absences of many of those volunteers recruited from southern communities who had a legal entitlement to take temporary leave from their regiments and return home just for the duration of the Autumn harvest.

The main plan behind the 1793 local military exercises was to defend an imaginary line that stretched between Brighton's parish church of St. Nicholas and the derelict church of St. Andrews, just to the east of the small village of Hove. To control the western flank, The Prince of Wales, who was the Colonel in Chief of the 10th Light Dragoons, set up his headquarters in St. Andrews' deserted cemetery. Whilst he was there, regimental aides saw to it that every effort was made to supply him with more than adequate accommodation which befitted his status. His personal tent for example, was far superior to anything ever seen on a battlefield for it had several ante-rooms and even a private kitchen. The Prince's initial enemy however was not the French, nor the weather or food, it was that the fields around his chosen headquarters were littered with heavy stones and once they were disturbed, the ground unleashed an infestation of earwigs. One of the very first battle orders the Prince gave to the 10th Light Dragoons was for his entire Regiment to set to and annihilate every one of his six legged foes.[3]

During the time the Prince was encamped at Hove, the National press produced frequent reports to illustrate how George took his responsibilities seriously and even informed the public that he often performed the duties of a captain of the guard, which included several night duties. However, in reality he was often known to have been lured by the more comfortable surroundings of his home in Brighton and given in to temptation by repeatedly absconding. This was particularly apparent on the night of 2nd September 1793, when a great storm arose and swept away many of the tents. Those who rushed to protect the Prince with concern for his safety need not have bothered for although theoretically on duty, he had departed much earlier for the luxury, warmth and overall comfort of his own apartments in the town.

It was whilst this first camp was in progress that a teacher called Vicesimus Knox was on holiday in Brighton. Being a Doctor of Divinity and Fellow of St. Johns' College, Oxford, he was called on to give the Sunday sermon at St. Nicholas Church, a task he unwillingly accepted. Although he made it clear to the entire congregation that this was his holiday and he would rather not speak again, they nevertheless requested him to give a further sermon on Sunday 18th August. Out of either bravado or an aspect of stupidity, he chose as his text, "Goodwill to all men" and "Love thine enemies." Such was his compassion, particularly with regard to accepting the point of view of those they appeared to have differences with, that a group of officers from the Surrey Militia thought he was making undue reference to the conflict with France and took exception to his presentation. They felt his sermon called them sinners for being at war with their fellow men and that he was undermining their determination for victory against Napoleon by advocating passivism. They therefore began to hound him throughout the town, shadowing him everywhere he went. They even followed him to the theatre and by a process of fixed stares and the passing of insulting notes, succeeded in getting the performance of, "The Agreeable Surprise" stopped and then encouraged all the audience to boo until he, his wife and their children left the theatre.[4]

On 14th October 1793, the camp between Brighton and Hove was finally struck and the volunteers were instructed to return home and the regular soldiers like those from the South Devon and East Norfolk Regiments began the journey back to their winter quarters. However, just as they were about to depart, a rumour reached the Duke of Richmond suggesting that a real attack by French Revolutionists, the Sans Culottes, was anticipated and he was obliged to send his Aide-de-Camps to recall every regiment. Packed baggage wagons had to be unloaded, neatly folded tents re-erected and the many carriages provided for the officers turned back. Sundry surplus equipment like wooden flooring and doors which had been disposed of to Brighton dealers for a fraction of their value was hastily repurchased at an exorbitant rate, showing those speculators an immediate handsome profit. A special rider was sent off to intercept the Royal Horse Artillery which had set off the previous Friday on their forced march to Woolwich, and they too were instructed to return immediately.[5] In the event, the scare of a French invasion turned out to be just another one of many false alarms.

The 1794 Encampment

By the Spring of 1794, Brighton had already become a very important venue for military training and this was commemorated by the minting of a special half-penny coin. The coin had the Prince of Wales' head on one side, and the date surmounted by his three ostrich feathers on the reverse. Around the milled edge were the words "Brighton Camp."

The new site chosen for the encampment was on the top of Brighton's Race Hill and it was here that the regiments practiced their battle routines, shooting at targets with both musket and field gun. The Race Hill was the highest point that overlooked the town and included a semaphore tower which connected with other stations along the coastline. It provided a vital communications link between Dover and Portsmouth and kept watch over the sea whilst also ensuring the accurate passing of naval intelligence. On a clear day, messages could be sent and received from the two adjacent stations at Shoreham and Seaford, or passed to ships that were at sea off the town. It was mainly used to give the Admiralty an advance warning of enemy activity in the Channel and to enable the naval authorities to direct their warships onto an intercepting course.

During the military training, there was always some speculation about an invasion, so when a rather garbled message was received to say that the French had actually landed in Brighton, it caused blind panic. In the confusion that ensued, the officers ordered every man to cease what they had been doing, grab a weapon and make their way immediately to the beach. As this rabble of soldiers in unfamiliar working uniforms suddenly appeared on the shore, the local fishermen and bathing women mistook them for the French landing force and grabbed anything to hand to defend their country against these presumed invaders. It eventually transpired that the message was just another false alarm which this time had been given by the Prince of Wales who was not testing the readiness of his soldiers, but purely saw it as a practical joke to which he was frequently prone. A prominent character artist of the day, J Nixon, captured the fight on the beach in an etching he entitled, "Brighton in a Bustle." It depicted the supposed French army complete with a guillotine, landing at Brighton and featured the distinctive rotund shape of the Brighton Dipper, Martha Gunn, leading the fisherfolk as they all defended the town against their own troops. Two caricatures of the well known generals, Sheridan and Fox, who were clearly privy to the Prince's jape are obviously amused as they watch the affray from the safety of separate bathing machines.[6]

"Brighton in a Bustle"

Other incidental military news of that year

Whilst the barracks at Brighton and Preston were still under construction, many of the horses used by the artillery units based at the Race Hill encampment and the bloodstock privately owned by the officers were put into private stables in the town. Unfortunately, on 16th of April 1794, the Bartholomews stables in the town square caught fire and although some of the horses were led to safety, nine troop horses were killed. Mr. Howell, the former owner of these stables, then moved to the new temporary army barracks in Unicorn Yard and provided new stabling for military horses and his own string of tradesmen's donkeys.

On the night of 17th October 1794, soldiers from the Prince of Wales' Own 10th Dragoon Guards spotted local smugglers unloading between 400-500 tubs (bottles?) of contraband gin onto Brighton beach. Not really acting in the public interest but more in anticipation of a reward, they used their side arms to arrest the smugglers. They then left two men to stand guard over the contraband spirit whilst they escorted the smugglers they caught to the local constable. On returning with customs officials who intended to deliver the load to Shoreham, they found that during the delay, their guard had rewarded themselves by taking the opportunity to consume a liberal quantity of the contraband and were falling about in a state of total intoxication.

The 1795 Mutiny

Although the tented encampments and manoeuvres around Brighton and its immediate environs continued to be a regular event, the conditions the volunteers were expected to endure and the rations they were given proved far from satisfactory. Local historian, Dudley Edwards' political pamphlet entitled *Soldiers in Revolt,* highlighted how gamblers arrived to exploit the soldiers during their little leisure time and how the more unscrupulous amongst the entrepreneurs benefited from the Napoleonic war by withholding fundamental foodstuffs in order to artificially inflate the prices. Profiteering raised the cost of survival even at the most basic levels and meant that the petty sums given to volunteers to support themselves and their families were grossly inadequate. Whilst the impoverished foot soldier had no money to send home and was confined to his quarters with hardly more than a 1/- per day before deductions (barely enough to purchase a quarter of a loaf or a quarter pound of beef) some of his officers by contrast were fast becoming carriage driving dandies, who drew their escorts from amongst the more elegant of Brighton's ever growing army of imported courtesans. It was only a matter of time therefore, before some soldiers rebelled against this injustice.

The privilege, if that was what it was, fell to the volunteers in the Royal Oxfordshire Militia who were then encamped at East Blatchington near Newhaven. Being unable to stand their deprivations any longer, they mutinied by raiding a local flour mill in the town and a grain ship moored in the harbour. Much of the food they liberated was distributed amongst themselves and some of the surplus was given freely to the poor in the community. Foolishly, for such an action broke civil law, the remainder was taken to Lewes and sold at more sensible prices to those being charged by the grain speculators. Incensed by the soldiers' indiscipline, the senior officers coordinating the volunteers' exercises sent a detail of men from their headquarters in Brighton to arrest the mutineers. However, being largely sympathetic to their cause, the arresting troop was repulsed without major incident.

Documents that recall the event do not outline whether those finally apprehended for the mutiny were just the ringleaders or whether the accused included other soldiers from the regiment who had been randomly selected to be put before the court martial being conducted in Brighton's Castle Inn. Whatever their status, each so called mutineer, was fully aware that public sympathy was with them and they intended to use that argument as the cornerstone of their defence. However, the presiding officers only had to consider if any man had broken military law by mutinying, and if that was proven to be the case, it was their duty to sentence each representative of the regiment to the severest punishment.

Even though the accused behaved with great dignity at their trial, they were nevertheless found guilty. Two of the men, Sykes and Sansom, who were known to have made a profit out of the insurrection, were convicted of breaking civil law and sent to Horsham gaol for execution. Six others were each ordered to receive 1,000 lashes and one man was sentenced to be transported to Botany Bay for ten years. As a deterrent to others who might consider challenging military authority in the future, Sergeant Edward Cooke and Private Henry Parish, who it is believed were selected to answer the charge of mutiny on the regiment's behalf, were sentenced to be publicly executed by a firing squad made up of their own comrades. Bravely accepting the inevitability of his fate, Sergeant Cooke was then marched out of the court and escorted to the Battery guard house in Artillery Place, where he was to be confined until his execution. During his brief stay there, he was frequently visited by the military chaplain as well as many well-wishers who offered him gifts of their precious food to make his last few days more pleasant. He also composed the following final letter to his brother.

Dear Brother,

I am sentenced to death and must die on Saturday, the 13th June; and I hope God Almighty will forgive me my sins. I never was nobody's foe but my own and that was in drinking and breaking the Sabbath, and that is a great sin. I have prayed night and day to the Almighty God to forgive me and take me to heaven, and hope my prayers be not in vain. I am going to die for what the Regiment done; and I am not afraid to meet death, for I have done no harm to no person, and that is a great comfort to me: there is a just God in heaven that knows I am going to suffer innocently.
Dear Brother,- I should be very glad to see you before I depart this life. I hope God Almighty will be a Guardian over you and all my relations, and hope we shall meet in heaven, where we shall be happy without end. So no more from the hand of your ever loving dying brother,

Edward Cooke.[7]

Sentence was due to be carried out on the 13th of June in the natural amphitheatre of Goldstone Bottom which was a traditional site used for the hanging of convicted felons. The lashings would be administered and the two men dispatched close to the very spot where in 1793, the tarred decaying body of the highwayman, James Rooke (later immortalised in Tennison's Rizpah) had remained encased in an iron gibbet for over a year. However, fearing an uprising by the comrades of the convicted men and the widely sympathetic general public, the punishment was postponed until the 14th so that responsibility for the maintenance of public order could be passed to the Prince of Wales' Own 10th Dragoons.

On the Sunday, many people from throughout the county gathered to watch the grim spectacle. Loaded cannon manned by the artillery were brought to bear on the assembled crowd, especially the Royal Oxfordshire Militia who were drawn up in front of thirteen other regiments to witness the punishments and thereby learn from the execution. The Prince's cavalry stood to the rear of the assembly ready to react if there was any disorder and twelve men all well known to Cooke and Parish stood with muskets ready to carry out the final sentence. The stage being set, as it were, the first six convicted men were then paraded through the ranks of their comrades before the sickening pageant commenced. Firstly, three of them were securely tied to the whipping frames and to the regular beat of a regimental drum the lashings started. After the three men had received 300 lashes each the officiating surgeon declared that further punishment would be fatal and the lashings mercifully ceased. Having been forced to observe each lash of agonising torment administered to their three colleagues, the three remaining men were pardoned.

Cooke and Parish were then ushered forward and escorted by the Reverend Dring, made their last winding journey through the ranks of their comrades who whispered messages of encouragement. They then took their places in front of their open coffins and received a final blessing. After hearing their confessions, the Reverend Dring, who was by now a close friend of the men turned to leave. Clearly affected by the injustice of the situation, he had asked the officer in charge if he could be some distance away when sentence was carried out. However, after some confusion over exactly where the two men should stand, the Reverend Dring had hardly gone a few yards before the accused were forced to kneel in front of their coffins and Cooke himself quickly gave the order for the squad to fire. It is said that the Reverend Dring was so shocked to hear the volley of shots ring out so soon, and when these were quickly followed by two more from the officer presiding over the execution, he collapsed, and subsequently died two days later.

As one last reminder of the consequence of their mutiny, each rank of soldiers from the Royal Oxfordshire Militia was then filed past the bleeding backs and the twisted corpses of their former colleagues. Before being finally dismissed from this scene of regimental discipline at its most cruel, the volunteers in the Royal Oxfordshire Militia were then informed that as a permanent reminder of their disgrace, they were to loose their "Royal" prefix and that in future their uniform would be changed from a Royal blue to a black and yellow.

The execution at Goldstone Bottom was captured by an unknown artist and copies of the resulting etching were quickly circulated.

The Execution of Cooke and Parish

For months afterwards, either morbid curiosity or tremendous public sympathy for the mutineers caused many people to come and visit the site where the execution had taken place. There they found an enterprising man had roped off an area where he regularly ensured the grass was still stained, and was for a few coppers, taking great pains to recall the public floggings and the execution in meticulous morbid detail. He graphically described how the use of shoddy coffins had resulted in the blood of Cooke and Parish congealing on the clothing of the burial party and was offering to sell especially manufactured bloodstained souvenirs.

The more radical in the community saw the military execution as an opportunity to stir up civil unrest and within days of the deaths, copies of the following inciting poem, acclaiming the men as martyrs to capitalist injustice, mysteriously appeared nailed to trees and pasted onto civic buildings throughout the county.

On those bloody numskulls, Pitt and George.
For since they can no longer send you to France
To be murdered like swine or pierced by the Lance
You are sent for by express to make speedy return
To be shot like a Crow, or Hanged in your Turn.

Ye Britons can you hear this and dauntless stand
Arise with thoughtful heads and nervous hands
Let Prudence guide, and let patriotic valour give the blow
And whelm in Ruin the Aristocratic Foe
Soldiers to arms, arise and avenge your Cause
You have of Chiefs and Patriots a matchless Train
To end a Pitt and bloody George's Reign.

Haste soldiers now, and with intrepid Hand
Grasp Sword and Gun to save thy native Land
For see your Comrades murdered, ye with resentments Swell
And join the Rage, the Aristocrat to Quell
Let undaunted Ardour each bold bosom Warm
To down with George and Pitt, and England call to Arms.[8]

Although few people could read, the poem, which fostered an allegiance with the French Revolutionaries was quickly learnt and relayed round Brighton's numerous alehouses. Fearing the matter could escalate into a possible insurrection, officialdom rapidly set about implementing major changes in military policy and conditions. Suddenly, much of the food being stockpiled by profiteers in order to maintain high prices was released and measures were finally taken to ensure that in the future, additional financial support would be available for the many thousands of families left behind by husbands and sons who had been recruited into the units of local volunteers.

The 1796 Encampment

The regular encampments at Brighton soon became an essential element of the social scene and continual references to them began to appear in art, music and literature.

Art

In May of 1796, the eminent artist, Francis Wheatley R.A. drew two etchings portraying the relationship between the military and the local people at Brighton Camp. The first of these actually entitled, "Encampment at Brighton," depicted three mounted troopers against a background of neat rows of tents and flying regimental pennants. Three troopers appear to be trading with a local farm girl who seems to have an aged father and numerous children to support. It may just be coincidence, but the socially aware might connect Wheatley's rather subtle emphasis on trading for readily available fresh provisions with the fundamental cause of the mutiny the previous year. Perhaps he is trying to indicate that only a year later, soldiers attending Brighton Camp had no difficulties with supplies and that the troubles of the past were as good as forgotten.

In his second etching, "Departure from Brighton" Wheatley dramatically used the sorrowful background of the gathering clouds in the winter's sky to portray remorse. Alongside the departing heavily laden baggage wagons and powerful cart horses he has illustrated the deep friendship established between a visiting soldier and a local family. A chaperoned girl's coy downward glance reflects the sadness she feels as the couple make their last fond farewells. Although clearly an indication of genuine emotion, the artist's romanticised point of view conflicts with the more general opinion being expressed in the media. Ordinary serving soldiers and volunteers attending the Brighton camps were considered to have been drawn from the lowest social order, few of which had firm plans to marry. Their motive in forming any relationship with local girls was commonly thought to be short term, insincere and purely physical in nature.

The Encampment at Brighton[9]

Departure from Brighton

Music and Dance

The military march entitled "Brighton Camp" penned in 1796, soon became a firm favourite for parades and the words were so catchy it was immediately re-christened, "The Girl I Left Behind Me." Not only did the tune provide a rousing regimental marching song incorporating glorious rhythms for fifes and drums, its popularity ensured that it was quickly choreographed into a folk dance, and traditionalists still perform it today.

"Brighton Camp" [10]

I'm lonesome since I crossed the hill, and over the moor and valley,
Such a heavy thought my heart do fill, since parting with my Sally.
I seek no more the fine and gay, for each does but remind me,
How swift the hours did pass away, with the girl I left behind me.
With the girl I left behind me.

Had I the art to sing her praise, with all the skill of Homer.
One only theme should fill my days, the charm of my true lover.
So let the night be e'er so dark, or e'er so wet and windy,
Kind heaven send me back again, to the girl I left behind me.
To the girl I left behind me.

Her golden hair in ringlets fair, her eyes like diamonds shining'
Her slender waist, with carriage chaste, they leave the swan repining.
Ye gods alone! Oh hear my prayer, to my beauteous fair to bind me,
and send me safely back again, to the girl I left behind me.
To the girl I left behind me.

The bee shall honey taste no more, the dove become a ranger,
The falling waves shall cease to roar, e'er I shall seek to change her.
The vows we registered above, shall ever cheer and bind me
In constancy to her I love, —The girl I left behind me.
The girl I left behind me.

Classical Literature

Jane Austen also made several references to the encampment at Brighton in her "Pride and Prejudice." The daughters of the Bennet family longed for their parents to take them to Brighton so that they could attend the monthly balls, imagining what it would be like to be courted by a whole regiment of militia.[11] Jane Austen highlighted how Lydia in particular romanticised about being invited to Brighton by Mrs Forster, the Colonel's wife. She wrote how openly enthusiastic Lydia was about promenading the streets of that;

> *"...gay bathing place covered with officers. She [Lydia] saw herself the object of attention to tens and to scores of them at present unknown. She saw all the glories of the camp —its tents stretched forth with beautious uniformity of lines, crowded with the young and gay, and dazzling with scarlet; and to complete the view, she saw herself seated beneath a tent, tenderly flirting with 6 officers at once."*[12]

The reality of the camp in Brighton was almost certainly somewhat different from the way Francis Wheatley illustrated it and Jane Austen's Lydia perceived it. Many of the officers, including their Colonel in Chief the Prince of Wales, were living well beyond their means and like Wickham in Jane's novel, were hopelessly in debt. As Clifford Musgrave discovered, a good number of officers had adopted a debauched lifestyle, seldom rising before midday and then spending all their time drinking, dining and attending the theatre with their paid escorts. Others were more pre-occupied with the bloodstock of their horses and the opportunities the town gave for numerous dalliances with local girls. Regardless of this, the impact of Jane's book when it was eventually published, continued to influence a substantial number of unmarried females to make their way to the town during the seasons when the Brighton encampment was at its height.

Even with this influx of eligible females, female company remained a valued commodity and officers as well as the rank and file often paid handsomely to acquire it. It was reported in the Gazette of 17th October 1796, that the town was inundated with an estimated 300 members of the Cyprian Corps (prostitutes) who were kept almost permanently busy throughout the entire duration of the encampment. The general feeling was that during their brief but regular seasonal stays in Brighton, many of these entrepreneurial ladies worked for and accumulated substantial fortunes.[13]

The 1798 Encampment

In 1798, the encampment on the outskirts of Brighton was situated in an area called Preston Hedges. It was to the north west of Preston Barracks, on Lord Pelham's estate, close to his country house in the village of Stanmer. Lord Pelham was a great supporter of the military and when manoeuvres were held, he frequently requested that the regiments made a marginal detour from their planned route from Lewes to Brighton so they would pass by his house. He would then have the opportunity to inspect the men and offer the senior officers refreshment. Furthermore, having the 1798 encampment actually on his land provided Lord Pelham with a ringside seat to observe the military exercises in their entirety and put him in a privileged position, for he could actually offer hospitality to the Prince of Wales.[14]

The Prince was present at one of the first exercises that took place on 27th September on the open ground of Newmarket Hill, and unfortunately, just like the earlier "Hove earwig" episode from 1793, this mock battle also proved to be yet another disaster. With the two opposing forces already drawn up in formation, the pre-arranged and meticulously planned sham fight should have given the Prince's defending army a glorious victory, but that was before another unforeseen enemy entered the affray and quickly caused both forces to retreat in total disarray. For, no sooner had the battle commenced when a huge waterspout came swirling in from the sea and headed straight for the battleground. Within seconds the skies opened and both armies were subjected to a torrential summer downpour. The officiating marshals had no option but to order both sides to cease fire. Although there was little cover, the battle field cleared in no time as all ranks from the Prince to the newest recruit scattered in a multitude of directions seeking immediate temporary shelter.

Once the storm had passed, the remnants of the two very bedraggled soaking armies re-emerged from beneath trees, bushes, horses and carts, most of which were some distance from the battle area. It then took some considerable sorting out to reform the troops into their correct positions in order to start the fight again. As planned, the Prince certainly won the day, but his prestigious uniform was saturated and his image ruined.[15]

The 1802 Encampment

The Public Record Office in Kew still has many of the original army pay lists and muster roles from the turn of the nineteenth century and these illustrate what soldiers earned each day before any deductions were made for their keep. Taking the volunteers from the Western Essex Militia as an example, when they were encamped at Brighton during the summer of 1802;

Sergeants earned 1s 6 ½d per day
Corporals earned 1s 21 ½d per day
Privates earned 1s 0 ½d per day
The young regimental drummers got nothing.

The only supplement to this rather meagre sum was an additional annual allowance which was awarded to enable volunteers to provide and maintain their own uniforms.[16]

The 1803 Encampment

Another fascinating mock battle took place upon Church Hill in Brighton on 25th September 1803. It was a planned sham fight between the militia of the South Gloucesters, The Sussex Volunteers, The South Hampshires and regular troops from the Flying Artillery.

The battle plan was for General Lennox's troops to attack Brighton and for Captain Bloomfield's artillery to repulse that advance. However, once the towns people got to hear that the battle was expected to take place only a short distance from their homes, they too assembled on Church Hill [Dyke Road] to watch the free show. J Bishop reported that by the early morning, every sort of transportation littered the battle area. There were barouches, landaus, landaulers, sociables, curricles, whiskies, fish carts, equestrians and pedestrians.

All went well whilst General Lennox's army began to advance firing the odd blank shot, but when Captain Bloomfield's defending force returned that fire using more powerful blanks in the field artillery, things began to deteriorate. Following the battle script, the attackers retreated into the area which was now occupied by the assembled mass of spectators and amidst all these onlookers, the planned fierce mock battle raged. Horses reared and bolted, ladies in all their finery fainted and the entire area became an enormous theatre with performers and audience intermixed. As usual the home forces won the day, but for many of the local inhabitants assembled there, it was their first experience of the actual mayhem of battle conditions.[17]

References

1	Sussex Archaeological Collections Volume 107			1969	pages 135-137
	Margary I D	*Militia Camps in Sussex 1793 and a Lady's Fan*			
2	Goodwin J E	*The Military Defence of West Sussex*		1985	page 69
3	Bishop J G	*The Brighton Pavilion and its Royal Association*		1903	page 17
4	Erridge J A	*The History of Brighthelmstone*		1862	pages 94-98
5		*The Times*	October	1793	
6	Musgrave C	*Life in Brighton*		1981	pages 112-113
7	Erridge J A	*The History of Brighthelmstone*		1862	pages 169-173
8	Edwards D	Political pamphlet *Soldiers in Revolt*			pages 18-19
9	Roberts W	*Francis Wheatley R.A, His Life and Works*		1910	
10	Scott H	*The English Song Book*		1925	Brighton Camp
11	Jane Austen	*Pride and Prejudice*	Chapter 39		
12	Jane Austen	*Pride and Prejudice*	Chapter 41		
13	Musgrave C	*Life in Brighton*		1981	page 111
	(Refers to The Gazette of 17 October 1796)				
14	E.S.R.O. ADD MSS 5440 No 259 Letters to Lord Pelham			1798	
15	Bishop J G	*The Brighton Pavilion and its Royal Associations*		1903	page 22
16	P. R. O. (Kew) Document WO 13 /700-725				
17	Bishop J G	*The Brighton Pavilion and its Royal Associations*		1903	
		The Battle of Church Hill	September	1803	page 31

LATER ENCAMPMENTS 1860-1872

In 1859, it was feared that the French were preparing for war again and as a safeguard against possible invasion, civilian volunteers were quickly recruited into the part time military reserve. They trained at drill halls all over the country and were committed to take part in major exercises most of which were held every Easter on the outskirts of Brighton. As in the past, Brighton's Race Hill became the most popular site for parades and frequent mock battles, all of which were designed to enable volunteers to develop familiarity with improved armaments and new military techniques. Not everything went according to plan however, for there were frequent mishaps and memorable incidences. In the 1861 Mock Battle for Ovingdean for example, when Lord Ranalagh deployed 15,000 troops in a mile long continuous line inland, a more realistic representation of battle conditions was accidentally created. Quite unexpectedly, one of the cannon in the flanking rows of artillery misfired, igniting the gorse and producing plumes of billowing white smoke which drifted across the battle-ground. Regardless of this, the home forces kept to their prepared script and continued to fire their carbines and carry out all the pre-arranged tactical advances and retirements, moving in and out of the smoke. With no soldiers detailed to play the part of an enemy however, the thousands of spectators who had an overview from the hillsides above, found themselves unable to understand the reasoning behind the all too frequent changes in the defending army's position.

The following year, in excess of 19,000 members of the Volunteer forces took part in similar exercises on Brighton's Whitehawk Down. Whilst some regiments boldly marched all the way from London, the majority arrived in Brighton as passengers on 16 special trains. Each special train carried a full complement of soldiers which was estimated to be 22 officers and 735 men. The rolling-stock consisted of a steam engine and its coal carrying tender, 22 passenger carriages and special trucks fitted out for the horses, guns and limbers of the field artillery. Further trains were provided at an excursion rate for those members of the volunteers' families who came to give their support and for other spectators who were attracted by the event. So great was the demand for transport, the South Coast Railway Company found themselves obliged to lease another 72 carriages from three rival railway networks. All went well and although a logistical nightmare, 6,922 troops were successfully transported down to Brighton from London Bridge Station in 2 hours 41 minutes. A further 5,170 took only 2 hours 20 minutes to arrive in Brighton from Victoria Station. Whilst this was being done, the railway continued to operate their scheduled services and many of these were used by a further 2,000 volunteers who travelled to Brighton from other locations in the southern counties. Witnessing the 1862 event, the author John Erridge, estimated that the local railway carried over 132,000 passengers on every day of that Easter weekend.[1]

Much of the organisation and administration needed to get the troops to the mock battle area was undertaken by the officers of the local Corps of Artillery and Rifle Volunteers who devised numerous alternative routes to get the soldiers through the narrow streets of the town. On their arrival at Brighton Station, the regiments and their associated military hardware initially marched or galloped down Queens Road with units then splitting off from the main column and wheeling down the side streets to pre-designated points for refreshment and regrouping. Those soldiers required to regroup at the Level or in the grounds of St. Peter's Church turned down Trafalgar Street, whilst those making their way to Steine Gardens turned at Gloucester Lane (now Gloucester Road) and North Road. Others due to assemble on the parade ground of Church Street Barracks or the gardens of the Royal Pavilion, made their way to the top of Church Street or North Street. In later years, the goods yard underneath

Brighton Station became a more preferred point for disembarkation and from then on Trafalgar Street provided the main point of access to the regrouping areas within town.

After the men had had a break for refreshment, on the stroke of eleven, a single cannon was fired from the lawns of the Royal Pavilion and this was followed by a chorus of regimental bugles each summoning soldiers and weaponry back into an orderly formation. The columns then set off past the cheering crowds to the Steine and made their way triumphantly along Marine Parade towards the County Hospital. At this point some soldiers broke ranks to prepare for the arduous task of man-handling their field guns and the more substantial pieces of coastal artillery up the steep slope to the top of the Race Hill. As these heavier guns were not provided with the steam driven traction engines like those which had been called into use at an earlier encampment in Dover, the limited strength of sleek army horses was augmented by powerful oxen and heavy shires from local farms, breweries and coal merchants. To manage these impressive but sometimes temperamental animals, soldiers in colourful uniforms often gave way to urban labourers in their Sunday best or rural farm hands in their bland country smocks.

Once at the top of the racecourse, the soldiers and horses re-took their positions in order to be inspected by senior officers and were then marched past an assembly of local dignitaries who stood proudly in the racecourse grandstand. Finally, with the prison-like buildings of the new workhouse as a backdrop behind them, the entire entourage marched off to the mock battle ground, passing the cheering crowds who had gathered either side of the wooden railings that enclosed the gallops of Brighton racecourse. *(See cover illustration)*

Whilst the regiments were then being deployed there were numerous side attractions. Shooting competitions were conducted and occasionally a hare was let loose for the spectators to chase. When the mock battles finally did start, some reporters seemed amazed that the volunteers demonstrated to everyone that they were fully conversant with all military strategies, especially forming squares against cavalry charges. J A Erridge made special note of a troop of volunteers from the Light Artillery who attempted a high speed charge towards the enemy using borrowed horses which they were unfamiliar with to pull their field guns. Considering the troop were only amateurs, he feared that there would be an accident, but amended his view when the troopers proved to be every bit as skillful as professional soldiers. If there had been any unfortunate injury, Colonel Hannington of the Sussex Volunteer Artillery and Mr. Burrows a local surgeon, had taken the precaution of establishing several first aid posts at key sites close to the race stand and on the western side of the mock battlefield. A field hospital was also erected in the grounds of Warren Farm, close to the Workhouse Industrial School.

References

1 Erridge J A *The History of Brighthelmstone* 1862 pages 288-293
Note: A more detailed study of these later encampments also appears in;
H J Woodard's M A Thesis, *The Volunteer Reviews of 1861-1879*

A Guide to a Collection of Illustrations

Whilst little physical or pictorial evidence of Brighton's early military encampments has survived, between 1860-1872, reporters from **The Illustrated London News (ILN)** were greatly interested in the regular Easter exercises which were frequently on the outskirts of the town. Their extensive reports covered these military manoeuvres in great depth and kept teams of artists fully employed making detailed sketches of all the events. Even if most are somewhat romantic, these illustrations reflect a favourable relationship between the military and the local community, although *"March from Brighton to the Assembly Point"* is perhaps an exception. This was drawn by an artist who sought to offer some comparison between the rich pageantry of the military processions with the tremendous poverty prevalent in Victorian Brighton. In many ways his more controversial view of the occasion makes a statement about social differences and contradicts the work of many of the other artists whose illustrations pandered to the more grandiose expectations of the more affluent amongst the paper purchasing public.

Troops Arriving at Brighton Station (ILN 23rd April 1870 page 429)

March from Brighton to the Assembly Point (ILN 30th April 1870 page 448)

The soldiers marching down from Brighton Station appear to be in Trafalgar Street which was just one of the numerous residential areas where Brighton's urban poor were densely compacted. Crowds are seen waving enthusiastically and an impoverished young mother has brought her fairly extensive family on to the streets to watch the free show. Other more fortunate children sit precariously on a plank suspended out of a high window. An inebriated man has left one of the many beer houses in the area and joined the procession, but not having a kilt, he has rolled his trousers up to the knees and dances a jig or hornpipe, which is the nearest thing he knows to a highland fling. Having heard rumours, a street urchin stands ready to dash in and peep under a soldier's kilt, whilst the unmended hole in the back of his own trousers informs his kid sister of his own lack of undergarments. Whilst everyone is focused on the parade, one tall and rather distinguished man overts his eyes from the festivities, not because he has little interest in the colourful occasion, but because he is a local police officer who has to constantly watch out for pick-pockets.

The Muster on the Level (ILN 29th April 1865 page 396)

Preparing for the Field (ILN 30th April 1870 page 440)

Man-handling the Field Guns up onto the Race Hill (ILN 13th April 1861 page 355)

Prince of Wales Reviewing the Troops (ILN 14th April 1866 page 349)

The Mock Defence of Ovingdean (30th April 1870 page 440)

The Mock Battle of Rottingdean (ILN 30th April 1870 page 445)

Amongst the illustrated crowd of spectators is the distinctive figure of, "Old Dad Moppet" from Rottingdean, a character that many people were familiar with. On important occasions like these, Dad Moppet always turned up dressed in his best boots and traditional rural smock, an outfit that became unique once he crowned it with a rather incongruous but nevertheless impressive, top hat.

The Middlesex Regiment Returning to London (Via Patcham)
(ILN 30th April 1870 page 441)

The supporting text with this illustration informs the reader that the soldiers were on their way to Brighton, but that was definitely not the case. Knowledge of the actual location some 3 miles outside Brighton in the small village of Patcham (now a suburb of the town) indicates that the regiment were leaving, not arriving. In all probability, the artist made a rough sketch and completed it later for he has removed the houses on the western aspect and used considerable foreshortening and romantic aesthetic sense to uplift the original Black Lion Inn from a less picturesque Georgian establishment further down the village, moving it into a more decorative flint faced Tudor cottage. Another change is the old tithe barn which for some reason has been repositioned in front of the church. Nevertheless the village is still clearly identifiable, especially the steep hill, the church's distinctive "Sussex cap" spire and the old village well which was later covered by a horse trough.

BRIGHTON'S COASTAL DEFENCES

THE TWO NEW COASTAL BATTERIES

Both of Brighton's new "Napoleonic" coastal batteries with their adjoining look out posts and military accommodations were built in 1793 to a specification laid down by Captain William Twiss of the Royal Engineers. Captain Twiss was also involved with designing the extensive chain of Martello Towers that were to be constructed along England's south eastern coastline. His idea was a unique contribution to national defence and ensured his rapid promotion to Major General.

Twiss was well aware that Brighton was already in the throws of rapid urban expansion and realised it would have been foolish to rebuild the new coastal batteries in either of their former positions, so he completely re-sited them. He believed that far better protection for the town would be achieved by repositioning the Eastern Battery on the edge of the low cliff opposite German Place (which was later renamed Madeira Place) and another battery on the western-most coastal extremity of the town, in an area which at that time was unnamed.[1] They would initially be armed with rather antiquated 20, 24 or 26 pounders (sources differ on this point), but in all probability they did include some of the original cannon that had formerly seen service on the old Eastern Battery which had been lost after being undermined by the sea. Several of these guns had in fact been salvaged from the sea and had since been laying unattended and open to the elements on the Steine. As recorded earlier, these guns proved to be far from satisfactory and although not officially condemned, most were certainly honeycombed with rust and considered unsafe. In the event of any enemy ship passing the shore during the crucial time these were Brighton's only coastal armament, many considered it more prudent to send a message to the encampment of the Light Artillery, who would willingly turn out with some of their more modest field guns in order to defend the town against a possible invasion.

Eventually, twelve heavy pieces of ordnance were supplied as replacement cannon, but these did not arrive in Brighton until 1801, when the initial phase of the first war against Napoleon was almost over. These guns were classified as 42 pounders, but were in reality modified French six and thirties, which had been captured by Lord Howe on the 1st June 1794.[2] The standard 36 pound shot they were expected to fire had to be altered to overcome certain peculiarities of the modified French barrel, which resulted in them discharging a projectile which weighed a fraction over 40 pounds. These adjustments certainly appear to have been worthwhile for John Erridge recorded that at midday on 17th August 1805, the gunners of Colonel Samuel Moore's Volunteer Corps of Sea Fencibles accurately propelled a shot a distance of 3 miles out to sea.[3]

The Napoleonic Eastern Battery. 1793-1807

Captain William Twiss' new Eastern Battery had perhaps one of the shortest periods of existence of all Brighton's coastal defences for it only lasted 14 years. It was sited on a piece of land which had been obtained on a 500 year lease from a Mrs Richards, who was to receive an annual rental of £15. The battery had a 65ft wide by 37ft deep platform built on the top of a 20ft thick retaining wall. Its initial armament was four of the town's original 20 pounders which were positioned behind a 20 inch high "en barbette" upright brick wall from where their range was expected to be in the region of 1350 yards. These guns were replaced in 1801 with the more accurate and longer ranged, French six and thirties mentioned earlier. However, just like its many predecessors, the gun emplacement could not stand up to the rigours of the sea and within a short while, a large proportion of the gun platform had fallen away. All that remained to support the battery was a foundation composed of loose stones and sand, which by 1807 had become so weakened by the explosions of the guns, that much of the site was in ruins. By the time the Eastern Battery was visited by H R Attree in 1809, its guns had been removed and only the Western Battery remained to defend the town against a possible invasion.[4]

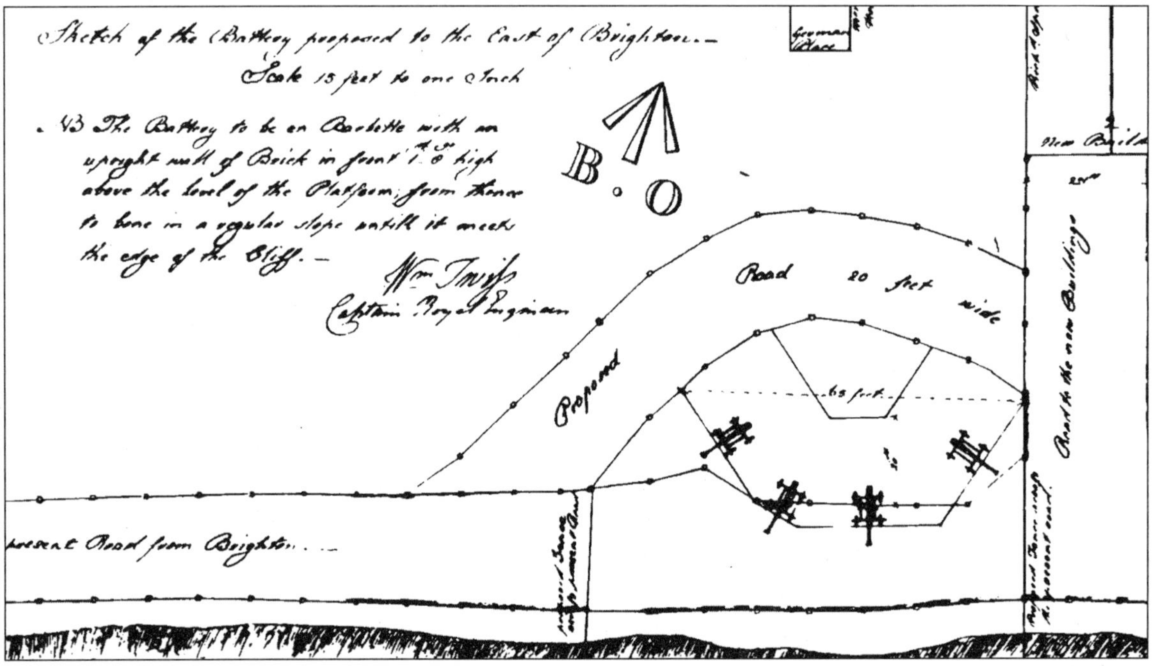

Plan of the New Eastern Battery (1793)

Later, when the entrance to the Chain Pier was built almost beneath the site of the former Eastern Battery in 1822, the executors of Mrs Richard's will began to prepare a case in an attempt to claim all the back rent for the ground. However, after much legal wrangling between 1827 & 1829, no payment was ever made and Mrs Richard's claim was eventually dropped.

An Artists Impression of one Battery in Action

The Napoleonic Western Battery. 1793-1858

The new battery on the western side of Brighton was built across several narrow strips of land which together measured some 116ft by 249ft. The individual plots, which cost a total of 200 guineas, had formerly been in the ownership of a Mr. Smith, Mr. Wilson, Mr. Whitchelo, Mr. Pocock and Mr. Pound, and their acquisition as a site for the Battery necessitated the demolition of a sawpit and workshop. Once built, the new Western Battery occupied a fenced off, stone paved enclosure and was positioned on the cliff opposite a parade of lodging houses which became known as Artillery Place. The Battery was a substantial structure, designed by Captain Twiss to include a small barracks for 26 men, a guard room, housing for the master gunner, a magazine and a furnace for heating the shot. It accommodated eight cannon which were again expected to be fired "en barbette" from behind a low upright wall. These eight cannon were also replaced by the French six and thirties in 1801, and afterwards in 1807, were further augmented by the four cannon which had been permanently withdrawn from the unsafe Eastern Battery.[5]

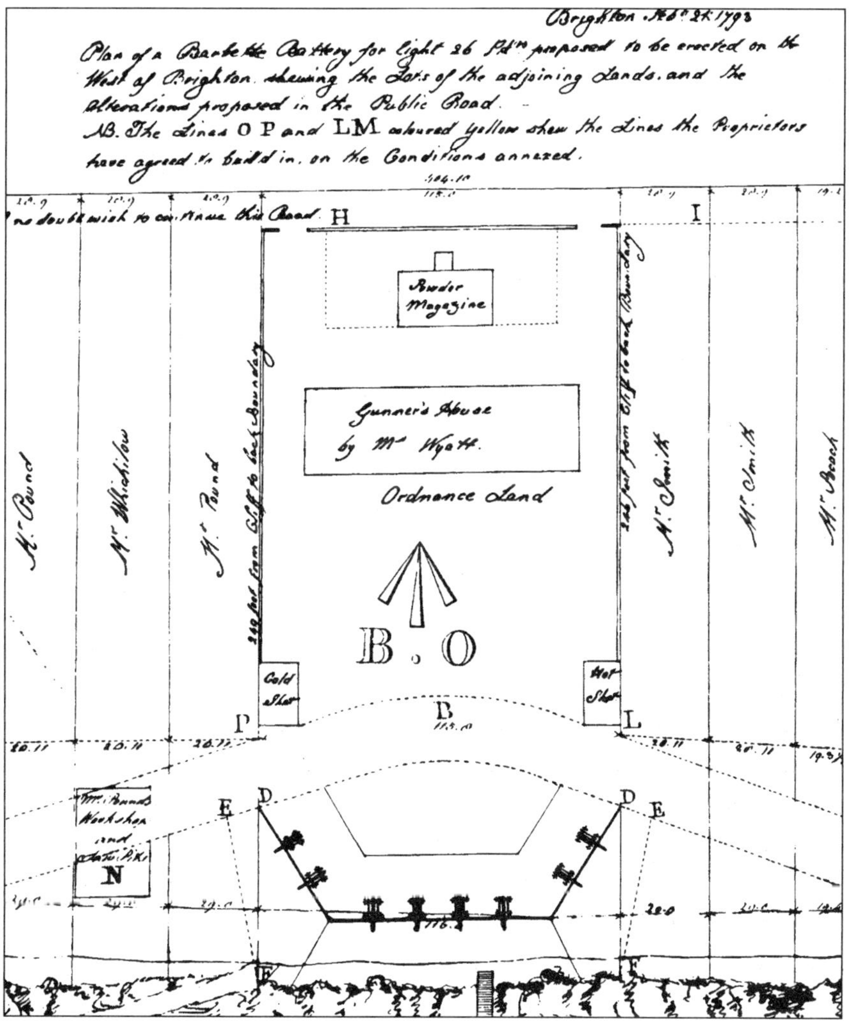

Plan of the New Western Battery: (1793)

Other available illustrations of the Western battery show no evidence of the upright "en barbette" wall and it is presumed that this aspect of the fortification was lost in 1829 when the cannon were moved forward to allow the coast road to pass behind them.[6]

The New Western Battery

The guns of the Western Battery only once opened up in anger when an English sloop called the Fly carelessly fired at some local smugglers and their shots landed close to the beach. Thinking the pursuers were French, the volunteer gunners on duty turned out and returned the fire, only stopping when the ships colours were eventually unfurled. By then the local smugglers had transferred their cargo of contraband to another vessel and made good their escape.[7]

The use of Brighton's coastal batteries during the second phase of Napoleonic wars was investigated in great detail by Lieutenant Colonel Laws of the Royal Artillery. He discovered that up until the end of the "Peace of Amiens" in May of 1803, both batteries had been the responsibility of the Corps of Sea Fencibles who were under the direction of Sir Edward Neagle of the Admiralty. With the renewal of hostilities against the French in 1803 however, overall command for the county's defence was passed to the Duke of Richmond who presumed that important anchorages like the one off Brighton would be very tempting for Napoleon. He therefore recommended Lord Hobart to find more volunteers to recruit into local coastal defence, especially men committed to essential occupations (like the fishermen in Brighton's substantial offshore fleet) offering to certify them as exempt from service in his majesty's navy or militia. As his offer was unconditional, within only a few weeks he had a list of 4 sergeants, 4 corporals, 2 drummers and 110 men, all of whom were Brighton fishermen said to have a working knowledge of "Great Guns." The four officers consisted of Lieutenant Colonel Samuel Moore (whose initial enlistment was in 1759 and who had been present at the siege of Gibraltar) supported by Lieutenant John Mills, Lieutenant Robert Dix and second Lieutenant Roger Wilsford. Their appointment greatly upset the Corps of Sea Fencibles for it drastically reduced the amount of time they could spend on shore in the relative comfort of the coastal batteries and meant that they had no alternaive but to return to their original role which involved patrolling the choppy sea off the local coastline in lightly armed gun boats.

Lieutenant Colonel Moore's newly recruited regiment was unofficially dubbed "The Prince of Wales' Royal Brighton Volunteer Artillery," but searches in the records seem to suggest that they just adopted this as an impressive title, rather than having it bestowed upon them. For them, image was of the utmost importance and these new volunteers modeled themselves and their uniforms on those worn by the Royal Regiment of Artillery, dressing in a Royal blue jacket and breeches, embellished with a red collar and cuffs and gold lace. Each man gave an undertaking to train unpaid every Sunday and received a shilling a day for each of the twenty other days of their annual commitment.

Once thoroughly trained, the volunteers had to complete a ten day spell of permanent duty on one of the town batteries. This training was somewhat limited, for practice with live ammunition for the guns cost 30/- a round and the administrative authorities always kept the units short of cash. Such were the financial constraints on the volunteers that it was a constant battle to find funds to pay for clerical services, stationery, postage, repairs to drums, fifes and swords, the provision of targets, boat hire, the routine maintenance and inspection of the guns, and the safe storage of essential equipment and ammunition. It was already known that official obstinacy over matters of finance had caused the Chichester Volunteers to resign *"en bloc"* so eventually on 17th June 1809, Lieutenant Colonel Moore wrote to the Secretary of State, Lord Hawksbury, to complain. In his letter he protested;

> *We only have 10 firelocks for which we are allowed 6 shillings and 8 pence per year for each to keep them in good repair, provided we can produce two Magistrates to inspect them. This is the only allowance we have to cover every contingent expense. Your Lordship must be well aware how very inadequate the allowance for clothing is. We are obliged to hire a store for arms, accoutrements, ammunition etc., and it costs at least 20 shillings every time we practice to moor the cask we fire at in front of the Batteries...*

Moore then indicated that many of the costs of running the Brighton Volunteers were coming out of his own pocket and suggested that if the situation continued, the town battery would be so poorly financed, salutes to acknowledge special occasions or the importance of visiting dignitaries like the Prince of Wales would have to be discontinued.

Unsatisfied with the response he received, Lieutenant Colonel Moore wrote again on 6th December 1809 and this time gained the support of Lieutenant General Charles Henry Somerset who was the Commander of all the Forces in Sussex. With a curious mixture of admiration and surprise, Somerset informed the Government that he knew the Brighton Corps were very efficient and attentive, and even though they were,

> *"...composed of the lowest order of tradesmen, who were liable to be called from their occupations, they still cheerfully submitted to this at the loss of their day's work."*

Somerset's recommendation was that the Government should relax their restrictions on allowances towards items such as uniform (£1 per year) and suggested that they should honour all the financial commitments they had first made when recruiting volunteers to defend the English coastline for the duration of the Napoleonic Wars.[8]

After the Napoleonic War had ended, Brighton's Western Battery was still kept in use for coastal defence and its smaller armament returned to being used for official salutes. Frederic Ogburn's reminiscences of his childhood in the 1840s recalled how he often visited the Battery where he estimated some 300 cannon shot were piled up in the rear. He was present during the experiments with early mines just off the Battery site and actually saw the John of Gaunt being blown completely in two by the force of a single explosion. He noted in his memoirs that some who were gathered there to watch this demonstration of new military weaponry, commented that the use of mines in warfare contradicted the accepted rules of conflict.[9] With a small boy's candour, Ogburn went on to document how 21 gun salutes were fired on Royal occasions and mused over this seemingly peculiar procedure. He recalled that only the small guns that lay flat on the grass were used and that the officer would first read something from a book before then touching a lit taper to a hole in the rear of the gun. Being unsecured, the recoil would often cause the cannon to leap several feet in the air and it would then roll over. Whilst it was being righted and re-primed, the officer would return to the book and read more from it. He explained that the gunners would rarely use the larger cannon because it was understood that the report from them would almost certainly break many of the windows in Kings Road. A common belief was that the noise might even be so great that the effect of it would be evident some 100 yards away at the top of West Street and in Western Road.[10]

In his research, the Architectural Historian, Anthony Dale discovered that the Town Commissioners unsuccessfully applied for the removal of the Western Battery in 1821 and again in 1823. On the second occasion, they even submitted draft proposals to The Board of Ordnance, to ask if they could move the Battery onto Marine Parade, some 230ft east of Russell House. Their request was again refused, this time on the grounds that the new site only offered a limited field of vision and once more would be too exposed to the ravages of the sea.[11] In later years, support to retain the battery came from no lesser person than the King himself, who in 1831, not only instructed Sir John Burgoyne to preserve it, surprisingly he considered it should be augmented by the construction of additional barracks.[12]

Years passed, monarchs changed, and Brighton's last remaining coastal battery finally outlived its usefulness, becoming just an inconvenience to the increasing traffic in Kings Road. So in 1858, almost 300 years since the fishermen of Brighton first volunteered to provide and man local sea defences, the Western Battery and its attendant buildings were removed to provide an apron for the Grand Hotel and enable the Kings Road to be widened.

No sooner had the Western Battery been removed, than it was feared that there was to be a conflict with France. Quickly many of the former members of the gun crews reformed and joined the Sussex Volunteer Artillery under the command of Lieutenant Colonel T Estridge. By then many of the local volunteers had moved on from deriving their income from the sea, and a substantial number were in the employ of local entrepreneurs. A contingent worked for Mr. Charles Smith Hannington, who along with three of his sons, had all become part time officers serving in the Artillery regiment. As can be seen later, Colonel Hannington played a substantial role in the furtherance of the local Corps by providing them with their first Drill Depot alongside the old Church Street Barracks and then by obtaining the former Eagle Iron Foundry in Gloucester Road, which was refurbished and became the permanent base for the Brighton Corps of The Sussex Volunteer Artillery.

There is still some speculation about what happened to the cannon that for years had lain on the grass of the battery like, "giant black slugs" pointing menacingly out to sea.[13] An illustration shown later suggests two may have been sent to Church Street Barracks for training purposes and another suggestion is that three of them were also taken up to Brighton Station to be converted into gate posts. An alternative suggestion is that these gate posts were formerly Russian guns captured during the Crimean War. However, an expert on military armament is of the opinion that whilst these gate posts do strongly resemble cannon, they could never have been safely fired and were just fabricated in the foundry of Brighton Railway Carriage works to look like cannon because at that particular time the Crimean War was in the forefront of people's minds and militaria was fashionable.

References

1 Goodwin J The Military Defence of West Sussex 1985 pages 58 & 60
2 Goodwin J The Military Defence of West Sussex 1985 page 59
3 Erridge J A The History of Brighthelmstone. 1862 page 71
4 Attree The Topography of Brighton 1809 page 62
5 Goodwin J The Military Defence of West Sussex 1985 pages 56 & 60
6 Goodwin J The Military Defence of West Sussex 1985 pages 62
7 Musgrave C Life in Brighton 1981 page 114
8 Lieutenant Colonel M E S Laws OBE, MC, RA (Retd.) FRHS *The Royal Artillery Journal* Vol. 84 October 1957
9 Ogburn F *Eighty Years of Reminiscences* 1914 page 11
10 Ogburn F *Eighty Years of Reminiscences* 1914 page 20
11 Dale A *Brighton Town, Brighton People* 1976 page 133
12 Goodwin J *The Military Defence of West Sussex* 1985 page 62
13 Kitchen F *Sussex History No 21* Brighton's Batteries May 1986 page 10

BRIGHTON'S TEMPORARY BARRACKS

When the war against Napoleon was declared, there was no official military accommodation in Brighton. Optimists thought the conflict would be relatively short and recommended the construction of temporary buildings which were only expected to be required for an extremely short period. In all, Brighton was to have four barracks and a military hospital, each being built of wood and constructed to a standard design used throughout the country. Ann Hudson, who conducted an extensive investigation into the numerous Napoleonic barrack sites throughout Sussex, found two articles in the 1793 Sussex Weekly Advertiser which both referred to the slow progress of huge rafts of timber for the barracks in Brighton as each made their way slowly down the river Adur.[1]

The war with France however, was not as brief as initially anticipated and local military strategy quickly changed with a decision to make the Infantry Barracks being built in Brighton's Church Street and the Cavalry Barracks in the nearby rural village of Preston, both permanent. This revision of the plans halted the further development of the temporary barracks under construction in West Street and Unicorn Yard, and the military hospital at the top of Church Street and sadly they soon became redundant. Nothing of the fabric of these three temporary buildings survives today, for at the end of the Napoleonic Wars, one had been sold off and the other two dismantled. All that remains as evidence that these buildings even existed comes from the local rate book and a few rather ambiguous references contained in material about Brighton's Regency past written by the prominent local Victorian historians, Melville[2], Martin, Erridge and Bishop. The general lack of contemporary information about these buildings and their occupants leads to a supposition that whilst officers were welcome, the presence of the ordinary common soldier in the highly fashionable Regency town of Brighton was considered a most unfortunate inconvenience.

WEST STREET BARRACKS

Unlike all of the other military buildings in Brighton, West Street Barracks do not feature in the 1814 town rate book[3] and the only contemporary information about them comes from H R Attree, who wrote in 1809 that these barracks could accommodate 132 soldiers.[4] Regrettably, that was the full extent of his coverage and nothing in the way of an illustration or local plan seems to have survived. Even the above mentioned Victorian historians, who should have had considerable access to oral accounts about West Street Barracks all seem to differ when it comes to actually pinpointing its exact location. For example even the very reliable John Erridge succeeded in placing the building on the two opposing sides of the same street when he wrote that;

> *"The barracks in West Street were at the corner of Little Russell Street, and afterwards became the Customs House."*[5]

His text created an impossibility, for early maps of the town clearly show Little Russell Street to be on the western side of the road whilst the Customs House was always on the east. Perhaps, in a temporary moment of madness, Erridge may have inadvertently sought to connect the West Street Barracks with Artillery Place and Cannon Street and in consequence had begun to confuse the barracks with the accommodation behind the battery which had been provided for the gunners of the coastal artillery. Another and more probable school of thought is that West Street barracks were indeed on the western side of the road, but sandwiched between the Royal George and the Kings Head. This site backed on to the West Street Brewery which was once owned by Henry Catt who also supplied military provisions and was occasionally known to have offered soldiers part time work during the winter. Erridge discovered that prior to returning to a tented encampment in the spring of 1794, soldiers from the militia who had been billeted in West Street barracks during the severe winter had found temporary jobs in the brewery's malt kilns. They were employed to heat up large blocks of stone, which were then taken up the road to the newly constructed Brighton Sauce Warehouse in Boyces Street, where the stone was being used to help the new brickwork and plaster dry out.[6]

UNICORN YARD BARRACKS (also known as Windsor Street or North Street Barracks)

Apart from having three different names, little can be established about the above barracks other than they were only used between 1793—1815 and were built next to the stables in the yard behind the Unicorn, an old inn which once was in North Street. They were sited on the piece of land now beneath the large Boots the Chemists department store near the town's clock tower. At the turn of the 19th century, this area abutted the former site of Brighton's most notorious slums called Petty France and Durham Place, which were built round an overflowing cesspool. Foot soldiers were probably the only individuals who were brave or rogue enough to go anywhere near this duck boarded, timber gantried, disease ridden ghetto, for they too had a bad reputation. A common belief was that apart from bouts of frequent drunkenness and time spent in adjacent doorways with the local whores, the soldiers here thought nothing of using their side arms for acts of intimidation or robbery and the bayonets they carried for forcing open windows and house breaking.

Access to the barracks was through the North Street entrance of the Unicorn, or from the Inn's yard in Windsor Street. Having no illustrations, it can only be surmised that the accommodation was very basic and the soldiers slept two to a bed in a number of bunks built into the corners of what was little more than a couple of timber faced ship-lapped huts. Similar buildings, designed and constructed to this specification were used all over England throughout the Napoleonic Wars. Brighton's 1814 rate book certainly dispels any idea that the buildings ever formed part of a substantial military structure, for it just describes them as *"Sundry Barricks"* sharing the inn's small yard with stables and a coach house, a warehouse and a granary, Mr. Holden's stable for donkeys, Mrs Cornwall's tripe shop and a number of other unimportant dilapidated tenement properties.[7]

It is assumed that the barracks in Unicorn Yard were originally constructed to openly circumnavigate the official order prohibiting troops and militia being billeted on licensed premises. The Unicorn's innkeeper clearly had overall control of the entire property and could even regulate the soldiers supplies of water which came from the inn's own 90ft deep well. This well, which re-emerged during site clearance in the early 1890s, was perhaps the only geographic feature of the Unicorn Yard Barracks to survive.[8] Unfortunately, shortly after its rediscovery, the well was refilled to enable the site development to continue and for reasons of public safety.

Perhaps it was the intimidating nature of the site, that resulted in the eminent local historian, H R Attree preferring to completely overlook the complex in 1808 during comprehensive research for his *Topography of Brighton.* Even at the end of the 19th century, it was only J G Bishop who reflected on the barracks, briefly likening the Inn's yard to Portland Yard which was adjacent to it and also indicating that the inn only really thrived whilst it had the patronage of the soldiers. He added that once the barracks were abandoned in 1815, the innkeeper acquired one of the former barrack rooms to use as additional stables[9] presumably in a futile attempt to regenerate some of his lost income. As stated before, it seems that all Brighton's nineteenth century historians considered the provision of facilities for the ordinary soldier a less desirable aspect of local development and most moderated their reports in case information detracted from the town's more important social image.

THE MILITARY HOSPITAL AT THE TOP OF CHURCH STREET

The military hospital appears in the town rate book as occupying what would appear to have been one of the most unhealthy sites in Brighton. Situated at the top of Church Street, the hospital was only a short distance from the town's overflowing cemetery and immediately surrounded by a series of stables, the local hog pound and Henry Tuppen's slaughter house.[10] Few local historians ever mentioned the hospital in their researches and it can only be assumed that the building was again of timber and conformed to the same traditional Napoleonic pattern being used at the Unicorn Yard Barracks, which was close by. Nothing is known about its construction or removal, all that has been recorded is that by 30th August 1825, the land on which it and its own adjoining cemetery once stood, had been engrossed into the newly completed Hanover Chapel.[11] In later years, Queens Road was constructed along the western aspect of the site to provide direct access from Brighton Station to the sea, and in the late 20th century, the chapel itself was removed. Today nothing remains of the former military hospital or the original cemetery for all have been engrossed into the grounds of the Brighthelm Centre which now occupies the entire site.

References

1	Sussex Weekly Advertiser		18.11.1793 & 30.12.1793
2	Melville L	*Brighton History, Follies, Fashions*	1909 page 89
3	The Brighton Rate Book	*An Equalisation of the Parish Houses, Stables, Coach Houses and other Buildings.*	1814 page 13
4	Attree G H	*Topography of Brighton*	1809 page 61
5	Erridge J A	*A History of Brighthelmstone*	1862 page 169
6	Erridge J A	*A History of Brighthelmstone*	1862 page 331-2
3	The Brighton Rate Book	*An Equalisation of the Parish Houses, Stables, Coach Houses and other Buildings.*	1814 page 13 item 108
8	Sussex Archaeological Collections Volume 39	*The Rediscovery of the Old Town Well.*	1894 page 213
9	Bishop J G	*A Peep into the Past*	1897 page 204
10	The Brighton Rate Book	*An Equalisation of the Parish Houses, Stables, Coach Houses and other Buildings.*	1814 page 37 item 34
11	Erridge J A	*A History of Brighthelmstone*	1862 page 106 & page 189

CHURCH STREET INFANTRY BARRACKS & THE ROYAL PAVILION STABLES

Introduction to Church Street Infantry Barracks

The substantial military complex built in the centre of Brighton which was commonly known as Church Street or Pavilion Barracks seems to have been totally neglected in almost every sense of the word. Both the military and local authorities kept few detailed records about the site and the buildings there were so badly maintained that eventually the entire complex had to be demolished. With the exception of some tell-tale marks still found on the tarmac of the Jubilee car park, nowadays little else remains to show where the barracks once were.

What scant information there is available in the local archives indicates that Church Street Barracks were originally expected to be temporary and that the men could occupy each building as soon as it was available. Whilst the inclement winters and Government restrictions on expenditure caused the soldiers to make undue haste and to compromise on materials, the continuous row of timber huts were not completed until almost 1796 for the Government realised that the conflict with Napoleon would be much longer and therefore, in mid construction, changed their military thinking to alter most of the buildings to make them permanent. The largely superficial upgrading consisted of improvements to the foundations, adding reinforcing brickwork and the later construction of a perimeter wall. Little else was done and Church Street Barracks remained largely in that format until 1869, by which time an almost total lack of periodic maintenance had turned the entire site into a public eyesore. In consequence, all the buildings were hastily demolished, allowing the local council to take over the southern area for the storage of their highways equipment and the North Street Slipper Baths to be built on the northern aspect. The latter, which became a nursery in 1976, still appears to occupy similar foundations to those which supported the barracks' accommodation block. Only one reminder of Church Street Barracks is left in situ today and that is the *"Barrack Yard"* plaque fixed to the wall of a once impressive arch which led to the North Road public baths. These baths were built in 1895 and were subsequently demolished in 1979. The arch now spans the access to a small car park that abuts the present Prince Regent Swimming Pool, a more recent facility built in 1981 to replace the two earlier public bathing amenities.

Although plans exist, extensive research has found no detailed illustrations of the Church Street site, and the Barracks can only be seen on an extensive roofscape of the housing around the Steine Gardens when they were surveyed from the top of the tower of the newly completed church of St. Peter. The picture itself is not very informative for all that can be deduced from this rather distant and obscure view from the 1840s, is that the barracks' main accommodation block was rather lengthy and had five brick built chimneys. Much of the remainder of the complex is hidden by the tall houses in Marlborough Place and numerous trees which cover the parade ground. Having trees growing in the barrack compound in an area which traditionally should have been an open space for drill and exercise is most unusual and this gives rise to some conjecture. It is doubted whether those trees actually existed and the artist may have introduced them in order to conceal a more unsavoury aspect of the military complex which had become apparent from his newly established view point. The use of greenery would have helped disguise the poor overall condition of the buildings and hide items such as punishment posts which had been erected in the barrack yard in order to enforce military discipline, The use of similar artifacts of correction were commonplace and can clearly be seen in more oblique illustrations which peek over the perimeter walls of the more extensive 19th century military complexes in Aldershot and Colchester.

Roof-scape Showing Church Street Infantry Barracks

Barrack Accommodation and Occupancy

With minor alterations, Church Street Barracks developed into a walled enclosure containing a row of timber clad huts on low brick foundations, alongside a range of barns, stables and a veterinary surgery. Some documents also list a military hospital being within the complex, but it is thought that London based officialdom may have mistakenly included the hospital which stood in the same street, but was in its own grounds some 100 yards away.

Although the barracks were originally intended as temporary accommodation for infantry soldiers only for the duration of the Napoleonic War, once the buildings were re-designated as a permanency, their flimsy construction proved the entire structure to be highly unsatisfactory for a large number of soldiers. The minimal improvements made to the accommodation were totally inadequate for the infantrymen who were told to share their billets with many troopers and artillerymen who were the over-spill from the far more spacious and prestigious Cavalry Barracks at Preston. It soon became clear that even though Brighton was a great attraction, a posting to Church Street Barracks was most unpopular and many soldiers sought to avoid it. Only H R Attree wrote with a modicum of approval of the barracks and after his visit in 1809, he reported that Church Street was a worthy structure, describing them as,

> *" ...most spacious, providing accommodation for 320 men, having an extensive yard and every suitable convenience."*[1]

For reasons best known to himself, Attree, whose descendants are believed to still have a property that adjoins the former barrack site, was one of the very few who presented a positive image of Church Street Barracks. All the other documents held in the national archives which relate to the site suggest that the entire complex was an expensive blot on the landscape, where conditions were both overcrowded and intolerable. For example, in 1806, three years prior to Attree's visit, these barracks were leased from a third party for what was then an exorbitant annual rental of £1314.8s.0d and although Attree may have only found 320 beds, the barracks were in fact the military quarters for between 836 and 892 soldiers.[2] The Infantrymen were so grossly overcrowded, that in common with many other barracks throughout the country, they and the many other soldiers residing there were sleeping more than two men to each bed. Ironically, as the following abstracts from various official returns reveal, even though barrack facilities were inadequate, conditions oppressive, and government expenditure on improvements and maintenance almost non existent, during the first thirteen years of the Napoleonic Wars the number of soldiers billeted in Church Street Barracks nevertheless continued to increase.

1794	80	Infantry	Barrack Master was Courtney Gidley.[3]
1804	670	Infantry	Barrack Master was Samuel Moore.[4]
1806	836	Infantry and a further 56 in the Hospital.	Barrack Master was Samuel Moore.

Returns made to the House of Commons after the end of the Napoleonic Wars, indicated that the number of troops billeted in the barracks was slowly being reduced to a more acceptable level. Sadly it is surmised that this reduction was not achieved by giving the soldiers greater personal space, but by the removal of the large hut which jutted out into the parade ground at a tangent to the main accommodation block.

House of Commons Returns of Barrack Accommodation in Brighton

January 1822	Barracks constructed for 0 Officers and 508 soldiers
	Occupied by 0 Officers and 508 soldiers[5]
March 1832	Barracks constructed for 0 Officers and 336 soldiers
	Occupied by 0 Officers and 234 soldiers[6]
1841 Census	Barracks constructed for 0 Officers and 336 soldiers
	Occupied by 1 Officer and 13 soldiers
March 1847	Barracks constructed for 0 Officers and 336 soldiers
	Occupied by 4 Officers and 92 soldiers[7]

With the sale of the Royal Pavilion and its stables to the town of Brighton, Church Street Barracks was left unoccupied in the belief that the more substantial barrack complex at Preston would be quite adequate for national defence. Within a decade however, fears of a renewed conflict with France and an awareness of the growing unrest amongst the Irish Fenians resulted in the decaying barracks at Church Street being temporarily re-commissioned.

1851 Census	Barracks unoccupied
1861 Census	Occupied by 0 Officers and 151 soldiers[8]
1863 (Government figures)	Occupied by 0 Officers and 272 soldiers[9]

Local guide books indicate that the only official entrance to the barracks was from opposite the Royal Pavilion Stables in Church Street, where the set of gates were guarded by an armed sentry. On the west of these gates were steps leading down to the basement of the Battery Sergeant's quarters and to the east were the regimental offices. Plans from 1823, 1838 and 1863 reproduced later, show that at some stage the regimental offices were given over to civil engineers and these subsequently became The Clerk of Works quarters and part of the regimental guard house. Behind these offices which formed the bottleneck of the main entrance was a small store and coal yard, opposite which was a further store with an ablution room which backed on to it. As can be seen from the following illustration, in 1837, military engineers altered the second of these stores in order to fit a new stove to it. The resulting document gives some idea of the general size and nature of the wooden huts which were then prevalent throughout the barracks.

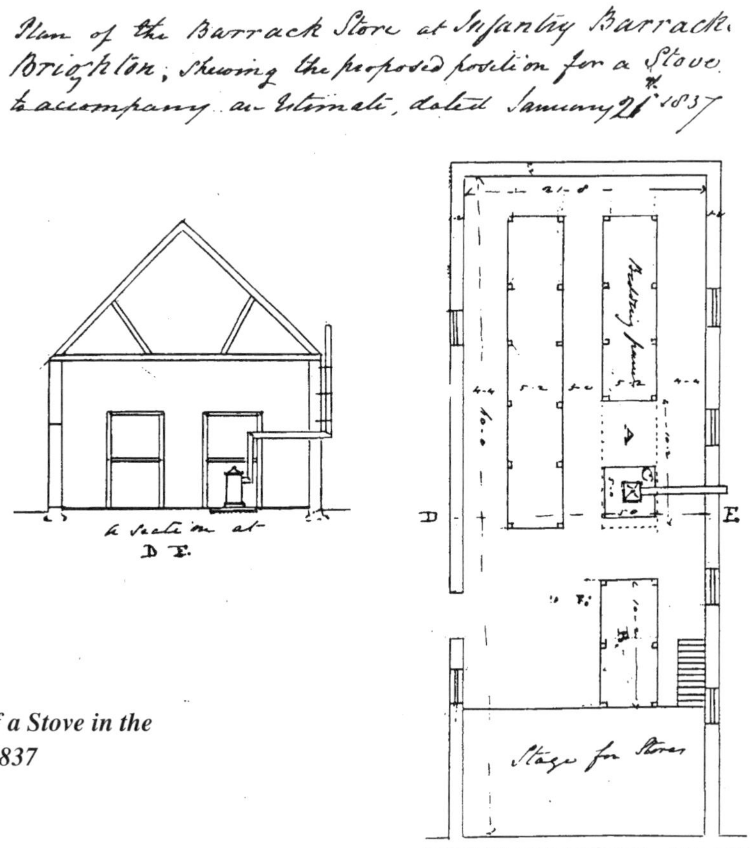

The Installation of a Stove in the Barrack Store in 1837

As numbers began to fall in the early 1830s, those Infantry soldiers still billeted in Church Street moved into the only remaining accommodation block on the north eastern side of the complex. This ten foot high building was constructed from a continuous row of the original huts, each being very similar to the illustration above. The entire block had a western facing verandah and stood hardly more than a pace away from the 7ft high wall on the barracks' eastern boundary. There were a total of eighteen rooms, all roughly measuring 30ft by 20ft, each with a fireplace, a ventilator and 3 or 4 windows. Assuming the rooms were fitted out to the format approved by army regulations, the four corners would have been permanently fitted with two tier bunks, making eight beds in all. Two men would have slept together in each bed, thus enabling sixteen soldiers to be accommodated in every room, whilst a seventeenth was out on guard duty. On the far northern wall (which it is believed had no official access to North Road) were the rifle store, gun room, armoury and the wash house. Alongside these was the barrack cookhouse which was fitted with 11 coppers.[10] Supplies of drinking water came from two 26½ ft deep wells *[superimposed on the 1823 barrack plan]* one of which was in the centre of the parade ground, some 41ft. from the accommodation block. In 1838, both these wells had to be deepened a further 8ft at a cost of £11.0.0½d because an increasing local demand for water was drastically lowering supplies from Brighton's subterranean springs. Further minor improvements were made in 1847 when the two wells were capped and replaced by hand pumps. More minor alterations appear to have taken place by 1863 and the water supply seems to have been diverted from the centre of the parade ground towards the western wall where two hand pumps supplied a series of stalls. Each stall's use is uncertain but they may have been for animals or just provided very basic washing or toilet facilities.

The Infantry Barracks Brighton (1823, 1838 & 1863)

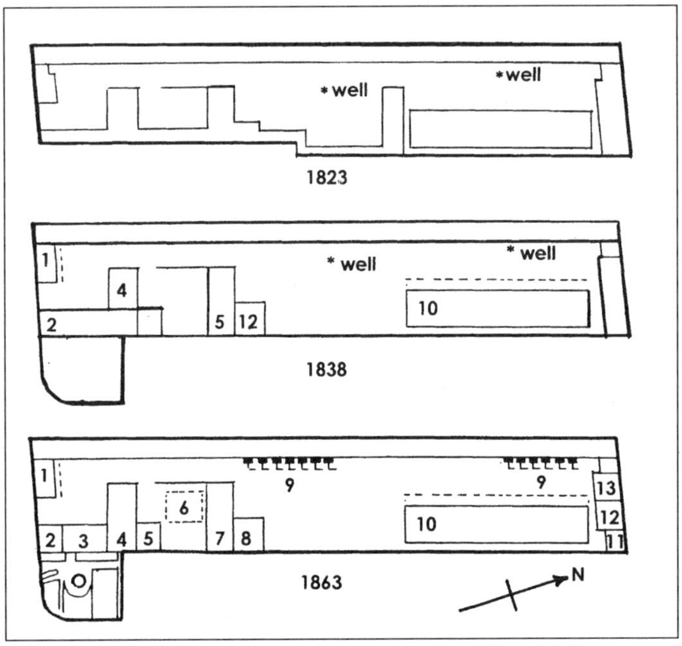

Key to the 1863 plan of Church Street Barracks

1. Battery Sergeant's Quarters
2. Barrack Offices (engineers' offices in 1838 plan)
3. Clerk of Works' Quarters (engineers' offices in 1838 plan)
4. Guard House
5. Barrack Store
6. Coal Yard
7. Barrack Store (see 1837 plan when a stove was fitted)
8. Ablution Room (a kitchen in the 1838 plan)
9. Stalls (washing or toilet facilities?)
10. Soldiers' Quarters
11. Wash House
12. Cook House
13. Rifle Store and Armoury.

Although the soldiers always anticipated that major improvements would at some stage take place, the above comparison between Piggot Smiths' town map of 1823, the well map of 1838 and the government's own 1863 plan of the barracks, all help illustrate how little was done in almost half a century. In all that time, only a little more land was acquired, one building was removed, the two wells were replaced by hand pumps and the several offices at the barrack entrance had been enlarged.

Marlborough Place

The plans of Church Street between 1823 and 1863 show a noticeable encroachment into the gardens of properties at the corner of Marlborough Place, providing an indication of how small additional strips of land were often considered worth acquiring. Transactions to obtain land adjacent to the barracks were frequently contemplated, but the complexity of local ownership, coupled with the almost inevitable delay factor of dealing with a government department, often resulted in the property being snapped up by a third party before details could be finalised. It was also common for the vendor to suddenly realise the high value of his land and by setting an exorbitant price for it, he was able to play the military, the local authority and any interested speculators, all off against one another. In November of 1834 for example, Southern Command's Lieut. Colonel Ellicombe wrote to the Barrack Master in Church Street to say that the present asking price of a submitted schedule of properties that he once could have obtained for £2,000, had risen in only a few months to £8,000. He highlighted the initial asking price of a local slaughterhouse which had formerly been available for £100, but had since escalated to £1,200. His letter, which referred to previous correspondence of 20th September 1834, acknowledged that whilst Church Street Barracks were certainly substandard, the proposed purchase of numerous freeholds from the multitude of owners with an interest in the patchwork of small strips of surrounding land was proving far too difficult. He went on to make it clear that the planned extension of the barracks would not be possible and any major expenditure in the way of improvements or additions to existing buildings would no longer be viable.

The only substantial addition to the site was the purchase of a quarter circle of land on the corner of Church Street and Marlborough Place, which between 1860-1869, was used by Colonel Hannington as a temporary site for an Artillery Volunteers' depot. Although unsubstantiated, the 1863 plan certainly seems to show a turning area for the limbers and several other familiar shapes which may represent the iron roofed regimental drill hall, the offices and a gun room. Nowadays the same site is occupied by an immaculate and well preserved late Victorian building which accommodates a local language school and a radio station. The present structure is an amazing contrast to the original buildings documented in 1834. In that year, the military acquired the land and many derelict buildings, including a former coal merchant's office and his yard, a corn-chandler's, an auction room, a tailor's shop, a beer shop, two sheds or stables and a dung pit, all for the bargain price of £60. These buildings were to be demolished and the anticipated income from the resulting sale of the rubble was to be used to finance the building of a 7ft. high x 225ft long wall to secure the barrack perimeter. This wall, which was completed in only a few months, had piers every 8-10ft and was topped with Portland coping. Unfortunately, the mortar used to bond the bricks in the wall was made of local chalk lime and sea sand, and like many other properties in the town, built down to a price rather than up to a specification. In consequence the salt in the sand soon washed out and the wall started to crumble away. This process of decay was encouraged by the soldiers in the barracks who found the new wall denied them access to the rear of the original King and Queen Inn, which backed on to the barrack yard. It is widely known that the soldiers and publican maintained a convenient hatchway cut into the wall in order to allow regular purchases of ale to be passed through into the barrack compound. This opening may be the same one that can be seen in the background of the illustration of two of the Artillery Volunteer's training cannon which is shown later.

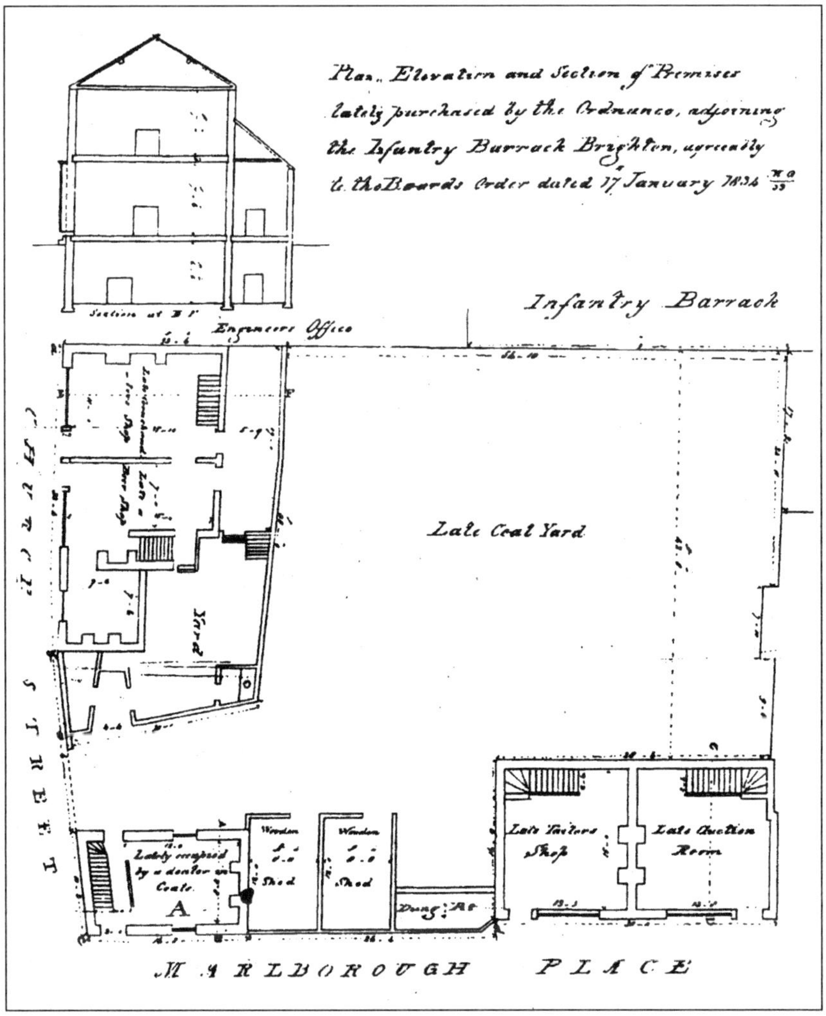

Former Properties at the Corner of Church Street and Marlborough Place (1834)

The Demise

As Church Street Barracks became less important, the buildings were allowed to further deteriorate until they became almost uninhabitable. By 1850, the Royal Pavilion and its adjoining stables had been sold to the town and as this transaction left the soldiers in the barracks with little work to do, Church Street Barracks were abandoned and the few remaining soldiers transferred to Preston. Only a renewed fear of trouble from France in the late 1850s brought Church Street Barracks back into temporary use and for a short period the uniformed sentry once again paced to and fro' from his box at the open gates. The reprieve was fortunately relatively short and in January 1869 the barracks finally closed for good, making way for the public slipper baths and council yard. Its removal must have brought gasps of relief from many serving soldiers as well as from a large number of the local residents of Brighton.

Looking back over the 74 years that Church Street Barracks were in existence, the military administrators had nothing to be proud of. The living conditions in these barracks were atrocious and the soldiers billeted there were often stressed or alternatively suffered from frequent bouts of depression. Little wonder that the appalling environment produced tremendous tension between many of the inmates, occasionally erupting into internal conflict, suicide and even murder. Trooper John Flood for instance, was consistently subjected to bullying by his comrade, John O'Dea, and when he could take the pressure no more, shot O'Dea in front of numerous witnesses who were standing just outside the guard room. This event, culminated in the court martial of trooper Flood and him being sentenced to execution in Lewes Prison. The full story of the murder in June of 1862, featured in the local paper, as did the trial at the Lewes County Assizes. Some years ago, the late John Montgomery made a rough draft of the reports, but his final version was never published. To put matters right, an adaptation of John's notes about the event that took place in Church Street Barracks during June 1862 is used here to retell the story.

Murder in Church Street Barracks[12]

In 1858 in the Irish district of Limerick, seventeen year old John Flood and nineteen year old John O'Dea, both enlisted in the 18th Hussars. Four years later, the Regiment had moved to Brighton where the majority of the men were accommodated in the showpiece Cavalry Barracks near Preston village. The over-spill of surplus troopers which included Flood and O'Dea, were redirected to the less congenial surroundings of Brighton's infamous Church Street Barracks. An unpopular feature of these barracks was that it had an unofficial and primitive system of courts martial which was contrary to army regulations, but officers and senior NCOs turned a blind eye to it. It involved the weaker men like Flood being illegally tried by their comrades for petty offences and having severe punishments inflicted on them. O'Dea, who was known to be a barrack room bully, manipulated this illegal system and enlisted the help of two other troopers named Filburn and Gassett, to become Flood's main persecutors. Together they used the courts martial system of barrack room terrorism to their own advantage and on one occasion, stripped Flood and lashed him with an army belt to such an extent that he openly confessed to the other soldiers that he wished he was dead. Flood nevertheless tolerated these chastisements, that is until 1st June 1862 when he was placed on sentry duty and O'Dea was detailed to clean all his stable equipment. Clearly aggrieved, O'Dea complained about the condition of the saddle and when Flood responded that it had been left clean, O'Dea countered with,

"You are a liar and I will have you tried by one of our courts martial when you come off guard!"

Flood remained on duty until seven, all the time brooding about a further beating. When relieved, he returned to the guardroom and didn't eat, instead he drank several pints of beer and two large glasses of rum. By nine o'clock he had resolved what had to be done and even though sentries were expected to carry unloaded weapons, he placed one of his ten cartridges into the breech of his weapon. There being no carbine racks in the guardroom, he then slumped down on the floor with his loaded gun beside him.

Less than fifteen minutes later, he was woken by sergeant Alfred Arter and sent back on duty guarding the barracks. Unfortunately, just at the very moment the duty sergeant left the guard in search of a light for his pipe, O'Dea was seen crossing the barrack square towards the guardroom. Stepping out from within the shadow of his sentry box, Flood issued the unusual challenge, *"Is that you O'Dea?"* On hearing him reply, Flood instantly rammed the carbine into O'Dea's stomach and after momentarily forcing him backwards in surprise, he jerked off his single shot. The round passed right through O'Dea, leaving his body just beneath his right shoulder blade and flattening itself like a shilling three foot up the western wall on the inside of the barrack gates. When the sergeant and the soldiers of the guard ran towards Flood, he made no effort to escape and just asked if O'Dea was dead. On hearing that he was still alive, he made a desperate attempt to reach the sergeant's sword but before he could do so, he was rapidly pinned down by the others. Despite efforts by the barrack surgeon to revive him, after lingering for several minutes, O'Dea died. Flood was then quickly handed over to the civil authorities who charged him with wilful murder and sent him for trial at the Sussex Summer Assizes in Lewes.

Throughout the trial, Flood sat almost motionless in a chair, clearly overcome with emotion and occasionally burying his face in a handkerchief. Many of his former comrades came forward to give him good character references, including the regimental Adjutant who affirmed that Flood was quiet and inoffensive whilst O'Dea had been quite the reverse. Further evidence was given by Corporal Brown, who told the judge more about the courts martial amongst the men, saying that he did not consider it his duty to inform the Commanding Officer for it was widely known that similar practices existed throughout the British Army, and that he himself had once been subjected to it. In Corporal Brown's opinion, there was certainly little point in mentioning it to the officers for they seldom took any interest in what any of the troopers did. Flood's council, Mr. Ribton, then made a powerful plea for mercy, pointing out that the young trooper had an excellent character, but claimed that he had been driven to this action after being continually subjected to brutality by some of his fellow soldiers. He believed that to hang an excellent soldier for such a crime would be a miscarriage of justice. He blamed the tragedy on both extreme provocation and on drink, which he said,

> *"...had maddened Flood's oversensitive mind."*

Flood's defence was that he had not been fully aware of what he was doing when he pulled the trigger and Mr Ribton requested that the jury should acquit Flood of murder and elect to find him guilty on the lesser charge of manslaughter.

Clearly public feeling was in Flood's favour, but whilst summing up, the judge reminded the jury that although they might consider recommending mercy, it was a prerogative only the Queen could exercise. In the judge's opinion it was certainly a crime of murder, not manslaughter, and the law should be particularly hard on a soldier who kills a comrade. The Jury then retired and returned after only 30 minutes with their verdict which pronounced Flood guilty. When invited to say why a sentence of death should not be passed, Flood sobbed,

> *"I was greatly excited at the time. I did not know what I was doing and I am very sorry for it."*

Unmoved by his plea, the judge donned the black cap and sentenced Flood to be hanged, causing Flood to collapse forwards in his chair, bury his face in his hands and weep profusely. Quickly the police officers in the court descended on Flood, bustled him from the dock and took him back down to the cells

After the trial many appeals were made to persons of influence and eventually the Secretary of State, Sir George Grey, pronounced that the Queen had agreed to extend her Royal Mercy to the soldier.

Sir George reflected on barrack room bullying by declaring that,

> *"This unhappy culprit* (had been) *goaded into an act by continuous and irritating provocation, from which he could not escape."*

Notice of Flood's reprieve only arrived at Lewes gaol a few hours before the execution and when the prison governor entered his cell, Flood's coffin had been made and even the metal plate with his name and the date had been fixed to it. He was informed that he had been saved from death, but was to spend the rest of his life in the harsh prison conditions of the time. Even so, when told he was not to go to the gallows and was instead condemned to a lifetime of penal servitude, Flood knelt down on the cold stone floor and thanked God for his deliverance.

Perhaps the Secretary of State was very aware of the intolerable conditions in Church Street Barracks, for O'Dea had not been the first soldier to meet an early death there. Earlier, on the 17th November 1819, a sergeant in the 19th Foot was also shot dead by a private in his own regiment. On this occasion however, the soldier faced a very swift trial and once found guilty, was quickly despatched to Horsham gaol for execution.[13]

In another incident, on 15th April 1822, a depressed soldier, Private Thomas Blamy of the 2nd Regiment of Foot, committed suicide by placing the muzzle of his musket into his mouth and a piece string from the trigger around his foot. His almost headless body was found in the darkness of the barrack yard by John Ireland, who believing Blamy to be just another drunken soldier, firstly kicked him in order to wake him and getting no response, lifted the headless corpse and tried to sneak it back into the barracks. It was not until Ireland reached the lit verandah that he realised that he was covered in copious quantities of blood and becoming fully aware of the morbid burden he was carrying, he let out a scream which alerted the sergeant of the guard.

At the inquest it emerged that Blamy had formerly been a servant to Lieutenant Jackson who lived in more pleasing accommodation outside the barracks, but the officer had dismissed Blamy after accusing him of stealing two shillings. Although no formal action was taken, for Blamy strongly denied the accusation, he was nevertheless reduced to the ranks and as another punishment, was returned to the barracks in Church Street. Being depressed by his situation and the inhospitable environment, he went on a heavy drinking spree in the old King and Queen and afterwards entered the Gloucester Tap where he met Mary Cross, a lady of ill repute. After spending some time in her company and leaving her with a black silk handkerchief as a reward for her services, he returned to the barracks for his musket and in the words of the Brighton Gazette, *"Promptly blew his head into atoms and scattered his brain over the walls of the immediate buildings."*[14]

Another soldier, Private Charles Millegan of the 2nd Battalion of the Coldstream Guards billeted at Church Street died on Christmas Eve 1835. In a probable festive stupour, he mounted the wall opposite Royal Crescent at a point where it used to drop steeply to the beach and plunged down the cliff. Devastated by the loss of a popular colleague, several of his comrades from the barracks provided a headstone which they then erected in the northern burial ground of St. Nicholas' Church. Millegan's elaborate epitaph read:

Oft may the tear his green sod steep,
And sacred be the soldier's sleep
Till time shall cease to run,
And ne'er beside his lonely grave
May Britons pass and fail to crave
A blessing on the fallen brave,
For such was Millegan.[15]

Civil Unrest

On one occasion, soldiers from Church Street Barracks were called out to help the local police quell a civil riot which had first started quite innocently when a group of over-enthusiastic revellers broke an ordnance originally introduced in Tudor times when many Brighton houses had thatched roofs. The act banned the uncontrolled use of fire within the town and meant celebrations to commemorate the gunpowder plot which involved letting off fireworks and parading a blazing tar barrel through the streets were prohibited. Things started to wrong on Tuesday 4th November 1817, when the town Principal, James Day and John Williams the High Constable, resurrected the act and instructed 16 police officers to arrest many of those who were lightheartedly celebrating. The detainments resulted in a mob standing in front of the Constable's house in Castle Square throwing stones and the town magistrate, Mr Runnington, trying to address the crowd but being shouted down. The High Constable's dinner guest, James White the local rate collector, (known as "The Soaker") ran to Church Street Barracks and instructed the guard of the 21st Fusiliers to turn out. Not knowing what to expect, the soldiers, who had only arrived in Brighton that day, fixed bayonets and marched off to the Steine where they regrouped in front of the magistrate. From his position of relative safety high on the steps behind the troops, the magistrate foolishly issued an ultimatum which prompted more stones and fireworks to be thrown. In his outrage, the Principal egged on by the rate collector, repeatedly ordered the Fusiliers to advance into the crowd, which they reluctantly did. In the ensuing melee which lasted until 2 in the morning, one woman was shot in the head, two Fusiliers were seriously injured by stones and a soldier succeeded in bayoneting a local peace officer, John Rowles, through the hip. It was reported that three inches of bayonet emerged on the other side of his body and that a few days later, Rowles died from the wound.[16] The event was captured in verse by the local poet, Thomas Herbert.

'...pretty soon he roars the rabble clear,
I'll read the riot act I've got here!
Go get the troops, and then we need not fear!
"How many troops?" the soaking hero cried
"All that you can," the learned bear replied
Now to the barracks flies the soaking chief,
and calls for troops; assuming bold and brief;
"I want some troops" to the Sergeant he did say,
"At your peril dare to keep away!"
The Sergeant Major to the guard house hies,
"Turn out the piquet there," he cries
The word is passed, the soldiers prompt obey,
And to the Steine that instant march away.

The troops arrive close to the learned bear,
he reads the proclamation in their rear,
and off he skulks, half dead with fear!
The soaking chief, like one bereaved of wits,
And almost going into fainting fits,
Charge on, he exclaimed, no mercy I'll afford!
Why don't you charge? You heard me give the word
Charge! Charge! Charge! Charge! the sabled knight cries
Charge! Charge! Again, the soaking hero replies.

The charge is made, alas poor luckless Rowles,
Thy life is gone through these ambitious fools.[17]

Married soldiers found being billeted in Church Street Barracks with their wives very difficult, but to some of the wives, life with a soldier was still a very attractive proposition. Widow Francis Short from Hangleton for example, took a new soldier husband who was adamant that he would not accept the additional financial burden of her fourteen year old daughter. So Francis' daughter Fanny (?) took that as an opportunity to marry Corporal Fudge from the Gloucestershire Militia. Unfortunately, by 1817 the Corporal had been killed, but quickly overcoming her distress, the young widow took up with two brothers who were again in Church Street Barracks, this time with the 51st Duke of York's Regiment of Foot. Sergeant Richard Harrap was the Regimental bugler and his brother Henry, a servant to an officer. Fanny Fudge soon married Richard, who rapidly found himself unable to adapt to the strains of being married whist still living in the environment of Church Street Barracks and promptly deserted. He was recaptured almost immediately and imprisoned. His release came after his seemingly impoverished young wife had by some undisclosed means, managed to acquire what was then the substantial sum of £40 and two replacement recruits, to buy him and his brother out of the army. The couple then lived happily together trading as shoemakers in Ivory Place until Richard died in 1828. Widowed for the second time, Richard's wife responded by marrying her former soldier brother in law, Henry. Later, Henry worked as a male masseur in Mohamed's Turkish Baths and the techniques he used were so successful that he and Fanny were able to extend their range of services and convert their small shoe shop in Grand Parade into a private massage parlour.[18]

Complex military relationships like the one above make the human element of historical investigations very difficult, for the military hierarchy from some of the regiments were frequently scattered throughout the town. When Lord Cardigan's 11th Hussars came to Brighton in 1840 for example, the soldiers were billeted in Preston and Church Street, privileged officers and their servants occupied the more comfortable surroundings of the Royal Pavilion Stables, and the extremely well off took on private housing for the duration of their stay. Lord Cardigan, who had his own private residence in Brunswick Square, found the appalling conditions in Church Street Barracks an ideal environment for punishment and when Lieutenant Forrest displeased him, he detailed the officer to relinquish his room at Preston and move to Church Street from where he should continue to carry out his duties. Knowing the conditions to be highly unsatisfactory, Forrest kept on his accommodation in Preston which resulted in him being brought before a panel of senior officers and being soundly disciplined.

The Local Volunteers

Before Church Street Barracks' eventual demise, its decaying facilities were extensively used by the Sussex Volunteer Artillery who were to become the last important group of semi-permanent residents there. Although discussed in greater depth later, the local volunteers were an essential part of the administration of the barracks between 1860-1869 as it was they who organised the traditional Easter encampments. In addition to providing the Volunteers with training facilities, the barracks' other main advantage was that its compound provided a convenient assembly place for military hardware and a marshalling yard for horses who at weekends were transformed from the drudge of their workaday role delivering supplies throughout the town and groomed to become majestic military mounts pulling powerful munitions. It was from Church Street Barracks that Henry Catt's elegant matched pairs of "Blacks" left for Brunswick Square pulling the field guns that were used to fire off the town's formal salute that respectfully marked the sad occasion of the Prince Consort's death. It was also an extremely well placed site to safely accommodate items such as personal armaments and heavy coastal artillery pieces. A rare illustration from inside Church Street barracks shows two such cannon mounted on heavy garrison carriages which had either come from Brighton's old Western Battery on the seafront, or were redundant munitions from a former battery in Eastbourne. These guns were used more as

n.b. It is not known whether the hatchway or door seen to the left of the cannon is the same opening that formerly led through the barrack wall to the bar of the old King and Queen.

The Two Training Cannon inside Church Street Barracks

training aids for it would have taken considerable planning and horsepower to re-site them both in a position where they could be effectively used for the defence of the town.

During their tour of duty at Church Street, The First Sussex Volunteer Artillery were once called upon to support the local police and patrol the streets because a possible Irish uprising was feared. The entry in Captain James (Bishop) Hannington's personal diary for 17th October 1867 reads;

> *"The Mayor has received an anonymous letter stating that the Armoury will be attacked at night and being the time of the Fenian scares it was deemed better that the hand of the police should be strengthened. Accordingly, I was put in charge of a strong detachment of men.*
>
> *At 11 p.m. I served out arms and ammunition and put a guard within the Armoury to keep watch quietly whilst I and my friend Mohamed, strolled casually about the streets and all the police were on alert.*
>
> *About 2 a.m. I noticed an unusual number of people about and four cabs came up and stopped near the expected scene of action. This looked suspicious and kept us all keenly excited.*
>
> *At 3 a.m. Mohamed spied a group of some dozen rough men lurking in the street. I walked up to them and passed through the middle of them. There were 10 in all and amongst them was an Armourer that I had discharged for misconduct. This looked very suspicious but I am afraid I was disappointed that nothing came of it. I suspect that the Armourer was at the bottom of the hoax."*

Conclusion

Although the former Infantry Barracks at the bottom of Church Street were originally conceived as temporary accommodation and only expected to last for the duration of the Napoleonic Wars, the barracks continued in use for nearly 74 years. In all that time, little attention was paid to maintenance and over the years the buildings steadily decayed providing an ever increasing depressing environment for all those billeted there. Strikingly the obituary of the complex was not written close to the date of its final demolition, but had already been published some 51 years earlier, at a time when many considered that the buildings had already outlived their usefulness. Quoting from his 1818 guide to Brighton, *The Brighton Ambulator,* C Wright issues the following brutally frank condemnation;

> ***"We wish it were in our power to speak well of this building but we regret to say that the situation is not adapted to the comfort of the military and it would be an improvement if this cumbrous pile of bricks and mortar and other nuisances were to be condemned and another building erected in a more conspicuous and agreeable spot."***[19]

THE ROYAL PAVILION STABLES

From the day the Royal Pavilion was first built it was always military policy to augment the facilities by installing some of the more privileged soldiers from the barracks opposite in the better accommodation over the stables. Many people were unaware of this and when the abandoned Church Street Barracks were recommissioned in 1859, John Erridge expressed his displeasure on their behalf. He was aggrieved that Brighton's prestigious and impressive Royal Stables were being used to house ordinary soldiers and wrote with a sense of humiliation that,

> *"Brighton had become a town with two ranges of Barracks which together were inadequate for the requirements of a single regiment,"*

He indignantly commented;

> *"Such is the reduced condition of England's military resources that she* (the nation) *is now compelled to beg house-room from the civil authorities to accommodate her soldiers."*[20]

Whilst Erridge's argument was indeed valid, his outrage did show some lack of understanding of the historic military ties that had been built up between Church Street Barracks and the Royal Palace and Stables. For the best part of 45 years the Royal Stables had provided both accommodation and gainful employment for soldiers living in the barracks. Erridge himself had clearly referred to this when he documented that in 1805, Promenade Grove had been purchased from the Prince's gardener, Mr Furner and a "new road" was being built across it by soldiers from the Royal Artillery. After its completion, he recalled that infantrymen from Church Street Barracks came under the direction of Mr. Porden and further helped in the construction of the Royal Pavilion. So grateful was the Prince for the men's involvement in the project, that on his birthday he gave each of them a guinea.[21] The soldiers then continued with the building work by assisting with the construction of the Royal Stables and its adjoining riding school, which was modeled on similar facilities built in Vienna. Although other plans for an indoor tennis court were never realised, the soldiers were still instrumental in completing the rest of the work which included a coach house, a farrier's shop and forge, the kitchens, 62 stalls for horses and accommodation for administrative staff. From then on it became a recognised duty for the soldiers billeted in the Infantry Barracks to have additional responsibilities like mounting a guard at key points of access to the Royal Palace and its grounds and performing other duties within the stables. Looking back at the evidence in the many early paintings and sketches by Nash, Good and Baynes which show the Church Street facade and the Royal gatehouses, it is evident that sentries from the barracks were always available to stand guard either side of the many entrances to the Royal Pavilion complex.

A War Office plan from 1863 further endorses that the Royal Stables were considered a barracks and shows that it included suitable accommodation for 6 officers and those NCOs who were their man-servants. By then the facilities had been improved to provide stabling for 82 horses, but excluded an area put aside for parochial offices, which is now occupied by today's library and museum facilities.[22]

After the closure of Church Street Barracks in 1869 there was a loss of a large pool of free manpower and the Royal Stables were modified to include many of the offices they have today. Erridge's comments in 1862 suggesting that the military had no right to occupy the building however, were very wrong, for from the very day that work started on the complex, the Stables and to some extent The Royal Pavilion itself, had always been an intrinsic part of the local military establishment.

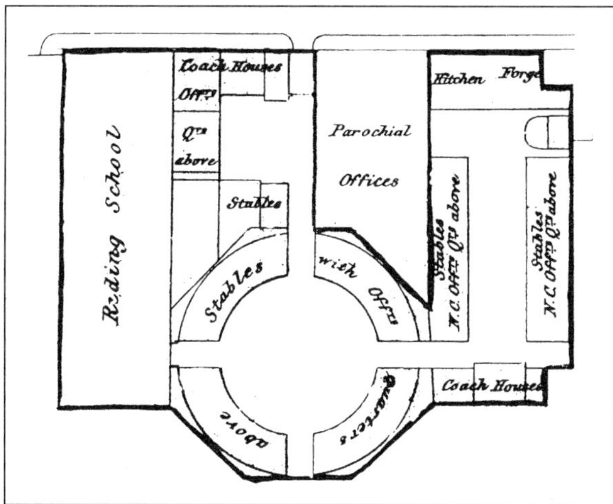

Left: ***Plan of the Barrack Accommodation in the Royal Pavilion Stables (1863)***

Below: ***Sentries at the North Gate. 1832***

References

1	Attree H R	*The Topography of Brighton*		1809 page 61
2	Public Records Office WO 40 25			
3	Public Records Office WO 40 07			
4	Public Records Office WO 01 88			
5	House of Commons	*Return of Barracks*	5th January	1822 (Public Records Office)
6	House of Commons	*Return of Barracks*	2nd March	1832 (Public Records Office)
7	House of Commons	*Return of Barracks*	31st March	1847 (Public Records Office)
8	Official Census 1841 roll 14 HO 107 1122 Folio 58 1861 RG 599 E D 43			
9	James H	*Plans of the Barracks etc., in the Brighton Division of the South Eastern District*		1863 (Public Records Office)
10	James H	*Plans of the Barracks etc., in the Brighton Division of the South Eastern District*		1863 (Public Records Office)
11	Brighton Gazette		12th June	1862
12	Montgomery J	*Murder in Church Street Barracks* Brighton Reference Library		Undated
13	Erridge J	*The History of Brighthelmstone*		1862 page 106
14	Brighton Gazette		16th April	1822
15	Erridge J	*The History of Brighthelmstone*		1862 page 244
16	Erridge J	*The History of Brighthelmstone*		1862 page 114
17	Erridge J	*The History of Brighthelmstone*		1862 page 274
18	Erridge J	*The History of Brighthelmstone*		1862 page 237
19	Wright C	*The Brighton Ambulator*		1818 page 125
20	Erridge J	*The History of Brighthelmstone*		1862 page 256
21	Erridge J	*The History of Brighthelmstone*		1862 page 268
22	James H	*Plans of the Barracks etc., in the Brighton Division of the South Eastern District*		1863 (Public Records Office)

DRILL HALLS & LOCAL VOLUNTEERS

To understand the need for Brighton's two nineteenth century drill halls, both of which now stand unused (in Gloucester Road and Church Street) it is necessary to gain a little more background knowledge about the colourful history of two units of local volunteers who occupied them.

The Formation of Brighton's Volunteer Forces

The Brighton Corps of The Sussex Volunteer Artillery and The Sussex Volunteer Rifles were both formed in 1859, at a time when the French were suggesting that the English had conspired with Felici Orsini to overthrow Napoleon III. The numerous political expressions of indignation being banded about at the time resulted in an element of the local population feeling threatened and led them to once again take the initiative by volunteering to train to repel a possible invasion. The government saw these units of local volunteers more as a joke than an efficient fighting force and as they had done in the past, initially withheld all support in the way of funding and accommodation, thereby making the volunteers subject to national ridicule. Fortunately for Brighton, several local businessmen and their ladies had more foresight and became benefactors by subsidising two of the units. Documents also intimate that further funding came through Lieutenant Colonel D'Albiac's contacts with Freemasonry. Every penny that could be obtained was important, for not only were the soldiers expected to provide their own uniforms, they also had to equip themselves with personal armaments. Eventually the government relented and made an annual payment of 10/- available to each soldier.

Brighton's two Corps of volunteers chose to dress in very similar uniforms which consisted of a basic tunic and trousers and a large peaked cap which was replaced by a forage cap after 1861. Around the waist the Volunteers wore a wide brown belt which was fitted with an ammunition pouch for the revolver each of them was allowed to carry and a frog to hold a 26 inch long sword. Soldiers in the Volunteer Artillery could be distinguished from those in the Volunteer Rifles by their cap badge which represented a grenade. Below is an early photograph of the uniform as worn by former Gunner W R Wood, who eventually rose through the ranks to become a Lieutenant Colonel in command of the Artillery Volunteers.

A review of the two corps written in 1896 stated that the colour of the volunteers' uniforms was originally Oxford grey, but earlier records from the archives of the Rifles dispute that. It would appear from documents of 8th February 1862, that the Brighton Corps of Volunteer Rifles had come into conflict with their own headquarters based in Worthing and were refusing to wear camouflage colours. They stated that greens and greys were the drab colours of the poor house and demanded to stay with the scarlet tunics they had originally kitted themselves out with. As they put it, they feared,

> *"...the loss of their individual image would also lose them recruits."*

Surprisingly, their demands were respected and for a while the Brighton Corps of the Sussex Volunteer Rifles

Gunner W R Wood of the Brighton Corps of The Sussex Volunteer Artillery

remained as the only rifle unit in the British Army to retain their original scarlet tunics rather than the more sensible regulation camouflage uniform which had become standard elsewhere.

The Brighton Corps of the Sussex Volunteer Artillery

When they were first formed, the Brighton Corps of the Sussex Volunteer Artillery had no official hall of their own and were obliged to drill with their unfamiliar "goose step" within the precincts of the Town Hall. However, after a short while they obtained permission from the government to utilise the small semi-circular piece of land just outside the Infantry Barracks, at the corner of Church Street and Marlborough Place. It was on this site that the Corps established their first artillery depot using funds donated by the Hannington family to enable them to erect a large drill shed and gun room which they had covered by an iron roof. This new base proved very popular and the strength of the local Volunteer Artillery increased so rapidly that several local employers began to take an interest and oversee their own batteries. Sixty of the volunteers were all existing employees of Messrs. Palmer and Green, and Battery number 4 was made up almost entirely from the employees from an expanding local department store owned by Mr. C E Hannington. His son, Charles Smith Hannington and three grandsons all joined as officers. Captain Silverthorne, who was the Adjutant, had battery number 5 and his recruits were drawn almost entirely from the local Friendly Societies. Captain Silverthorne also tried to raise a mounted company of Horse Artillery similar to those recruited into the "Honourable Artillery Company" but abandoned it when he found the expense too great. By 1869, when C S Hannington had risen to the rank of Honorary Colonel, he was able to report that Brighton's Corps of Volunteer Artillery consisted of 560 men, 514 of whom were continually active. Each artillery man had to be over 5 feet 6 inches in height and to be classified as "active" needed to complete a minimum of 30 drills in his first year. The acceptable height for those recruits enlisted into the Volunteer Rifles was set slightly lower at 5 feet 3 inches.

As well as being expected to provide themselves with uniforms and accommodation, the Artillery also had to obtain heavy armament to drill with. This setback was initially overcome by the manufacture of a wooden gun made exactly to the same measurements as an 18 pounder, which the men affectionately christened "The Quaker" on account of its inability to shoot anything. Once they had their permanent depot, they secured two iron 24 pounder guns and carriages, each weighing well over two and a half tons from the redundant garrison in Eastbourne. The M P for Brighton, Mr. James Wright, further augmented the local Volunteers' armament by contacting the Duke of Cambridge and organising the temporary loan of four field guns from the Militia stationed in Lewes. These guns complemented the old 40 pounder practice cannon which had also been permanently mounted within the yard of Church Street Barracks. In 1869 the Artillery Volunteers were to relocate to a new depot in Gloucester Road and after having well outlived their usefulness the antique 40 pounders were replaced in the late 1880s by four more practical 16 pounders.

The Volunteer Artillery's first commanding officer was Lieutenant Colonel J Estridge who was succeeded in 1863 by Lieutenant Colonel G C D'Albiac. D'Albiac died in July 1867 and although there were many Masonic brethren at his funeral, only Sergeant Major Menzies attended in uniform. His replacement was Colonel T H Johnson, who retired in August of 1868. Command of the Volunteer Artillery then passed to Colonel Charles Smith Hannington, who was surprised to discover that the government intended to sell off the Church Street Infantry Barracks including the plot of land on which the Artillery Depot stood. Being forced to look around for an alternative site, he fortunately discovered that the former Eagle Foundry in Gloucester Road was available. On 4th January 1869 he informed his Corps of Artillerymen that he had acquired the building and they would be moving in to it once it had been refurbished as a drill hall. The refurbishment was to include an officers' mess and rest rooms and similar facilities for the sergeants and NCOs. The gunners were also to have a large recreation room, but a restriction was to be placed on their smoking. When finished, the main hall measured 106 feet by 36 feet and was connected to a smaller hall which was annexed to the gun sheds and harness room. Initially the complex also included a canteen and a house for the Sergeant Major, but when Colonel Hannington retired because of ill health in March 1872, Colonel Tester had these removed. Although the frontage of this Artillery Drill Hall still remains in situ, just recently the inside has been totally converted for residential use.

The Gloucester Road Drill Hall

With leadership of the Volunteer Artillery frequently changing, it is easy to overlook the considerable part played by Lieutenant Colonel Charles Smith Hannington in providing permanent accommodation for the Corps. Mr. Michael Hannington, a direct descendent of Charles provided a brief résumé of the Colonel's history, in many instances using extracts taken from the diaries of Charles' son James, who became a Lieutenant in Brighton's Artillery Volunteers at 16, a Captain at 17and later became a Bishop. "Bishop" James Hannington wrote that his father Charles entered his grandfather's retailing business in 1826 which he subsequently inherited in 1865. During his early years, Charles served under his father as a volunteer on Brighton's Coastal Battery and after that closed in 1859, he transferred over to the newly formed Brighton Corps of the Sussex Volunteer Artillery. Charles introduced his sons, Samuel, Philip and James into the Corps which he eventually took command of in 1868. He did a great deal for the advancement of the local volunteers and even after his retirement in March 1872, the regimental band would still go uninvited to play outside his house in Hurstpierpoint. It therefore came as no surprise when his services to the local volunteers were recognised, and in February 1873 he was officially given the title of Honorary Colonel in Chief of the First Sussex Regiment of Volunteer Artillery.

Charles Smith Hannington, whose portrait is shown below, proudly wearing his military uniform, died in Hurstpierpoint on 7th June 1881.

My thanks to Michael Hannington for allowing the reproduction of a photograph of Charles' portrait.

Lieutenant Colonel Charles Smith Hannington

The earliest available military photograph of Charles Smith Hannington was taken between 1864-1865 when he was still a Major. It included Colonels Estridge, D'Albiac, and Johnston, all of whom were to precede Charles as Commanders of the First Sussex Artillery Volunteers. Charles' three sons also appear in the same illustration which from left to right the officers are; Lt. Col. D'Albiac, Captain P Hannington, Sgt Major Menzies, Captain S Hannington, Major T Hayter Johnston, Major C S Hannington, Lt. Col. Estridge, Lt. James Hannington

When it came to competitive shooting, both the Brighton Volunteer Artillery and their associates in the Volunteer Rifle Corps proved to be highly efficient so the photograph on the right provides an interesting contrast with another kindly loaned by Mr. A E Tuppen, shown below. This was again taken in 1865, but shows the men in Sergeant George Tuppen's victorious gunnery crew from the Brighton Artillery Volunteers at Shoeburyness. In that year the team won first prize, which was a cup presented by Queen Victoria.

Officers of the 1st Sussex Artillery Volunteers (1865)

The gun crew consisted of; Sergeant G Tuppen, Corporal Lewer, Bombardiers T Chatfield and David, and Gunners, Hobden, Row, Hill, Whitehead, Dudman and Coppard.

The Brighton Gun Crew from the 1st Sussex Volunteer Artillery (1865)

Note: *The original cup that the Brighton gun crew won is still treasured by the Tuppen family*

In the eight years that followed Colonel Hannington's retirement, command of the Artillery Volunteers passed rapidly between Colonel J Siverthorne, Major Lane, Colonel F A Weatherly and Lt. Colonel J Jefferies. Colonel H Tester took over in March 1880 and retired on 17th July 1886, putting the former gunner, Colonel W R Wood in command until he also retired in August of 1889. He was succeed by Colonel E N Edwards, who in November 1892 passed the responsibility on to Colonel T Chatfield. In July 1896, command of the Artillery Volunteers finally came to Colonel C G Boxall. Colonel Boxall was an, 'innovative ideas man' for it was he who first carried out experiments with a purpose built armoured train believing it to have potential within the British Army. Working with Mr. Billington, the Superintendent of the Brighton Locomotive Works, he designed a rotating armoured shield through which a 40 pound breach loading gun protruded. Its advantage over previous designs which were nothing more than ordinary 40 pounders mounted onto railway trucks, was that it could revolve on its axis in any direction. Its first trials were at Newhaven on 19th May 1894 where the gun proved to be a great success.

The Armoured Train

The War Office favoured the Colonel Boxall's ideas and by the end of the century had made considerable funding available to the Brighton Locomotive Works for the gun's manufacture and maintenance.

The Brighton Corps of the Sussex Volunteer Rifles

Brighton's Corps of Volunteer Rifles were formed at the same time as their Artillery counterparts, but for many years were less fortunate, for whilst they were very efficiently administered from a small orderly room in Princes Place, they had no official drill hall of their own. Although actually formed in November of 1859, no official parades took place until June of 1860 when the 300 men who had enlisted assembled on Lindfield Common. Once the Brighton Watch Committee became fully aware of the large numbers of volunteers, they quickly cancelled the Corps' planned church parade destined to take place in Brighton on the following Sunday, fearing the sight of so many armed soldiers might lead to a breach of the peace. Perhaps as some sort of recompense, the Town Hall was then made available for most of the Volunteer Rifle Corps' drills and they were held there until the late 1880s. On the occasions that the Town Hall was not available, drills shifted to the Corn Exchange, but as their rifles were still stored in the Town Hall's armoury, this proved to be a highly unsatisfactory arrangement. Later in the 19th century the Corps began to drill at the Corn Exchange on a more permanent basis until they took over their new drill hall at the top of Church Street.

As with the Volunteer Artillery, several of the early recruits worked their way through the ranks to become officers. For instance, the first man to enlist, private Jenner left the Corps as a Captain, and when private, F Hallett joined, he proved to be such a good shot that he was sent to the school of musketry in Hythe and returned as an Ensign. The Corps' success in many shooting competitions owed much to Hallett's skills as both an instructor and competitor, and by the time he retired, he was a Major. Initially, recruits were armed with muzzle loading Enfields, but these were converted to Snider rifles in 1860, a weapon which remained in service with the Corps until 1885, until it was replaced by the Martini Henri. To be efficient, each recruit had to show a reasonable degree of accuracy when firing 15 rounds of Morris Tube ammunition, followed by a further 42 rounds of ball. Few had any difficulty in obtaining high scores and the volunteers' rifle team often put many regular soldiers to shame. The Volunteers' clerk's meticulous records are also a joy to read, for almost every competitive success was celebrated with a lavish dinner, most of which are described in minute detail. Their shooting practice was greatly helped by the Corps having ownership of the ranges at Sheepcote Valley which they had acquired from a Mr. Beard. The site was very popular for the Easter military exercises, and shooting competitions always formed an intrinsic part of the training programme. The site was also a particular favourite with a local man called Chitty and his son who were both able to make a substantial income from illegally digging the lead balls and other shot out of the protective embankment. Eventually the two men were arrested and charged with wilful damage by ruining the site. Each was sentenced to 1 month in prison, with the alternative of a fine of £1.16s.0d. Considering the same site is now officially part of a Council tip with local traders recycling domestic refuse under licence, in the last 100 years, nothing much in this area has changed!

The early chain of command of the Brighton Corps of Volunteer Rifles is captured in the following copies of two photographs taken in 1862 and 1873. The first shows Lieutenant Colonel Moorson, outside the Corps' offices with several of his brother officers. Soon after this picture was taken Colonel Moorson passed responsibility for the Corps over to Captain Woodhead who quickly handed it to Captain Roupell. In 1863, Major H A Crofton took charge and after only 3 years passed the command to Lieutenant Colonel Penton. Fortunately, Lieutenant Colonel Penton was to remain in charge until 1879, before he in turn passed the command over to Major Henry Verral, who eventually retired in 1884 with the rank of Lieutenant Colonel. Colonel Verral, was to return to the Corps for a brief spell after the untimely death of the charismatic officer who replaced him, Lieutenant Colonel Tamplin.

Above: Officers of the 1st Sussex Volunteer Rifles in 1862

Below: The Officers of the 1st Sussex Rifles in 1873

Although the officers appeared to change quite frequently, many of the Volunteer soldiers spent a substantial number of years with the Corps. Amongst them was Colour Sergeant Edwin Gillam who originally joined the Brighton Corps of the Sussex Volunteer Rifles within a year of them being formed. He and many of his senior colleagues completed over 35 years of service, an achievement which was acknowledged in 1893 with an award for their long service and loyalty.

Colour Sergeant Edwin Gillam and other Senior NCOs from the 1st Volunteer Battalion of the Royal Sussex Regt.

(My thanks to Mr. M J Gillam for the loan of the above photograph)

Lieutenant Colonel Tamplin

It was under the leadership of Lieutenant Colonel Tamplin that the Brighton Corps of Volunteer Rifles really prospered, for soon after he took charge in 1884, the Brighton Corps increased in strength to 724 men and grew from 6 to 8 companies, which soon became known as the First Volunteer Battalion of the Royal Sussex Regiment. Furthermore, Colonel Tamplin immediately began to rectify the appalling situation that the men had endured by having no permanent base in which to drill or practice. Being a wealthy owner of a substantial local brewery, he had adequate funding to negotiate the purchase of a substantial piece of land at the top of Church Street close to the former site of the Napoleonic military hospital and to finance the design and construction of a brand new purpose built drill hall which he leased back to the Corps at a peppercorn rent. As can be seen from the following series of photos taken in 1896, the building which was completed in 1890, was quite impressive, having an extensive purpose built drill hall, a large canteen, and provided comfortable messing facilities for the officers, sergeants and men.

The Church Street Drill Hall in 1896 and 1996

1. *The Drill Hall Today*
2. *Inside the Drill Hall in 1896*
3. *Canteen in 1896*
4. *Officers Mess in 1896*
5. *Sergeants Mess in 1896*

Sadly, Colonel Tamplin's ill health brought about his early retirement and he regrettably died in 1893. He was buried with full military honours, having a funeral procession led by the regimental band, flanked by the officers and men of both the First Volunteer Battalion of the Royal Sussex Regiment and the local Artillery Volunteers. Further support came from many regular troops and great numbers of ordinary citizens who over the years had grown to respect Colonel Tamplin and who all turned out to give him a proper military send off. Close to his coffin would have been his former Colour Sergeant, Edwin Gillam, who is shown below in the full dress uniform of the First Volunteer Battalion of the Royal Sussex Regiment.

Colour Sergeant Edwin Gillam

(My thanks to Mr. M J Gillam for the loan of his family photographs)

Like their counterparts in the Volunteer Artillery, the Brighton Corps of Volunteer Rifles were not without innovative ideas. Sadly they could not produce anything as impressive as an armoured train, but by the turn of the century they had introduced signallers and a company of stretcher bearers into their regiment and became the first unit in the British Army to use cyclists for scouting purposes.

Reference Note:
Unless otherwise stated, much of the research material for the text and illustrations in this chapter has come from the membership of "The Sussex Family History Group," and the August/September/October editions of Views and Reviews published in 1896.

PRESTON CAVALRY BARRACKS

Preston Cavalry Barracks circa 1850. *(Either by De La Motte or C Wing)*

Introduction to Preston Cavalry Barracks

Having one substantial barracks in Brighton and another only a few miles away in the then relatively unknown rural village of Preston, has led to some confusion in early documents. Prior to the 19th century, the village of Preston was considered a separate community and as such the Brighton guides paid little attention to it. Some news reports in the 1795 & 1796 *Sussex Weekly Advertiser* did occasionally make brief references to the barracks at Preston, but added little of consequence and just advised their readership that it was hardly more than a collection of weather boarded buildings, built on a permanent brick foundation.[1] However, as soon as Preston Barracks showed signs of rapidly outstripping Church Street in military importance, the media and officialdom quickly granted the complex unofficial status as a suburb of Brighton and the name "Preston" was often dropped. From then on, almost all official documentation distinguished between the two military establishments by referring to them as either Brighton's Cavalry or Brighton's Infantry Barracks.

Apart from being an overnight transit camp for soldiers who were arriving in and leaving from Brighton, Newhaven and Shoreham during the first of the Napoleonic wars, the extensive facilities at Preston Barracks were initially underused. By the turn of the century, its 33 ship-lapped huts[2] still only housed 215 horses and a total of 169 soldiers, most of whom were serving in The Royal Horse Artillery or The Prince of Wales' Own Regiment of Light Dragoon Guards.[3] Little is known about the men, for although Preston's ecclesiastical census of 1801 saw fit to record their wives, children and camp followers who all added to the village population, no detailed listing was ever made of the regular or volunteer forces because parish officials were instructed to disregard them. By then the Barracks' canteen had been re-designated as a hospital and soon almost all of the temporary wooden huts were to be replaced by permanent brick buildings coated in a sandy white lime. Mr Henry Attree, clearly did not favour the continual upgrading at Preston, for his local guide of 1809 revealed that he had personal preference for Church Street where he possibly had a financial interest. Admittedly he found the barracks acceptable, but otherwise he just dismissed them as being:

> *"Immensely expensive buildings north of the Bear Inn and toll gate, which offered the expected standard of accommodation to the Horse, Foot and Artillery Regiments billeted there."*[4]

That same gate across the road was still taking a toll from travellers to Lewes some 58 years later.

1867 Postcard showing the Toll Gate outside Preston Barracks

Most other Brighton guide books followed Attree's line by ignoring the existence of Preston Cavalry Barracks throughout the Napoleonic wars, for it was 1818 before the very first comprehensive description appeared in the *Brighton Ambulator.* A more polished version of this text quickly followed in Wallis' *New Brighton Guide* and then, with certain minor alterations and updates, many of the other local guide books began to plagiarise these two versions.[5] The following compilation of all those descriptions, the majority published between the late 1820s to 1830s, provides perhaps the most informative and complimentary impression of the Cavalry Barracks in Preston ever to be written in the pre-Victorian period.

> *"Preston Cavalry Barracks is situated on the Lewes Road about a mile from Brighton in a healthy valley and possesses the advantages of both air and prospect. It is a handsome pile of yellow brick buildings which were erected in 1795 and have a frontage that occupies a space of 500 yards from south to north. The Barracks are calculated to contain over 625 private soldiers and non-commissioned officers.*
>
> *The south wing contains 22 rooms used as officers' apartments, together with rooms for their servants, the mess, the Barrack Master's office, the riding school and the kitchens etc., beneath which are the coal cellars. A forage barn is situated centrally. The north wing is composed of stables with 30 rooms for the soldiers above them. Five wells span the westerly aspect of the three buildings. The military kitchens and other offices are detached, which adds much to the comfort of the men.*
>
> *The barrack hospital is a neat and commodious building and is situated with its own independent well on the northern edge of the complex. It is together with an armourer's shop, a tailors, a smith and farrier's, a shoemaker's, a saddler's shop and a school room. The guard room and magazine are on the southern extremity of the barrack yard.*

In 1801, stables with reception spaces for 400 horses were erected in front of the barracks, extending along the Lewes Road. These accommodations were removed in 1818 and a lofty wall built which completely shut the barracks off from public view. Some of the finest Artillery and Cavalry Regiments in the kingdom are normally stationed in the barracks and their splendid bands frequently play on Steine to the delight of inhabitants and visitors.

As the following sketch map shows, many of these original buildings were still being used in 1863.

A Simplified Sketch of Preston Cavalry Barracks. (Adapted from a War Office plan produced in 1863)

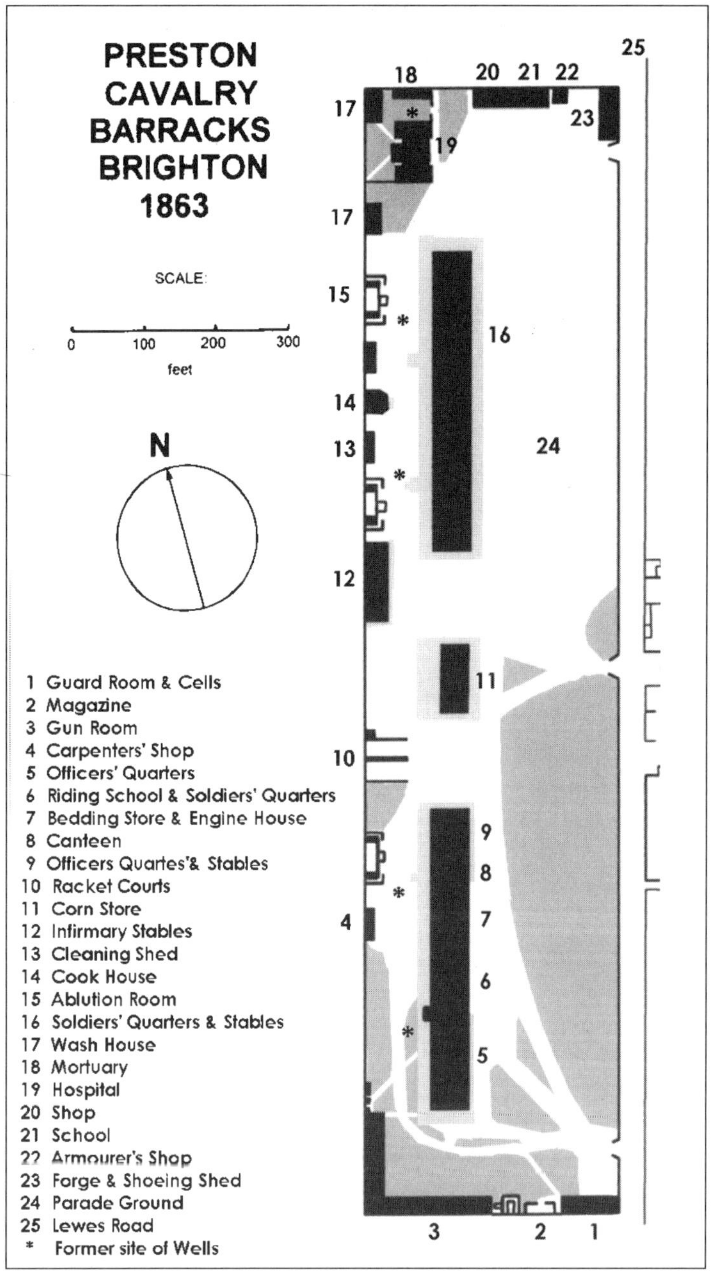

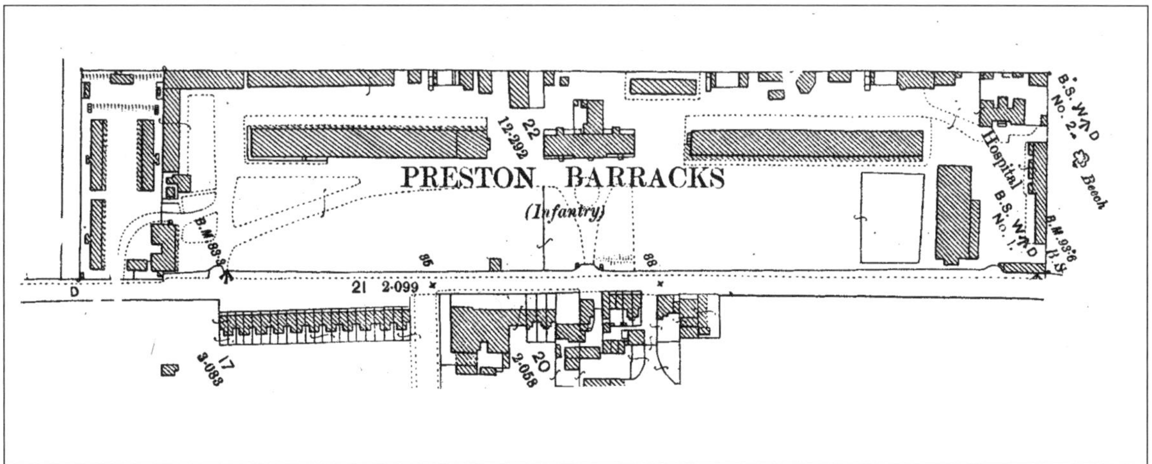

Ordnance Survey of Preston Cavalry Barracks in 1898

The Crimean Block

All the early descriptions of Preston Barracks were made well before the southern boundaries were extended onto an additional rectangular piece of land. As can be seen from the 1898 ordnance survey map, that compound had a further four buildings on it, all of them added during the Crimean War. These same buildings also featured in outline on a map produced in the late 1850s which was drawn to show the full extent of the Hollingdean Waterworks.[6] Unlike the rest of the barracks, these blocks traversed the compound and may have first been erected as accommodation for civilian personnel, or as shown in a post Victorian barrack plan from 1912, were an early attempt to provide separate married quarters for the more senior of the non-commissioned officers. For a short time two of these buildings fortunately survived the initial demolition of the barracks, but sadly their reprieve was only temporary and eventually they too disappeared and were replaced by a modern development. Now all there is left to remember the very distinctive military style of these buildings built during the Crimean War are a few photographs taken just before their demolition. One of the less impressive of these was so covered in dull remorseful shadows, that it needed to be sketched for the details to be revealed.

Sketch from a photograph of one of the Crimean Barrack Blocks

Accommodation

Although figures are conflicting, in the early years, the peace time role of Preston Cavalry Barracks was to provide accommodation for around 200 soldiers and 230 horses. However, as Anne Hudson found in her more detailed research into Napoleonic Barracks throughout Sussex, in time of necessity, adjustments and structural alterations were continually being made to accommodate more troops and horses. A random collection of official publications about Preston Barracks seems to bear this out.

Preston Barracks Accommodation from Early War Office Papers[7]

July	1794	162 Cavalry (Barrack Master William Randall)			
February	1800	169 Cavalry & 215 Horses (Barrack Master Samuel Moore)			
September	1801	An agreement made for new temporary stabling for 1,200 horses to be built and additional stabling for a further 100 horses to be hired in Spring Walk[8]			
	1802	Constructed for	26 Officers	679 Soldiers	? Horses
	1802	Occupied by	18 Officers	169 Soldiers	215 Horses
	1804	Occupied by	60 Officers	679 Soldiers	715 Horses
	1804	The Barrack Hospital	Occupied by a further 84 patients		
	1806	Occupied by	26 Officers	679 Soldiers	215 Horses
	1806	The Barrack Hospital Occupied by a further 53 patients			

Preston Barracks Accommodation from House of Commons Returns[9]

25 March	1812	Budget for improving accommodation and stables £26,300			
5 January	1822	Constructed for	26 Officers	196 Soldiers	255 Horses
		Occupied by	23 Officers	250 Soldiers	296 Horses
2 March	1832	Constructed for	26 Officers	196 Soldiers	255 Horses
		Occupied by	18 Officers	221 Soldiers	110 Horses
31 March	1847	Constructed for	26 Officers	196 Soldiers	255 Horses
		Occupied by	11 Officers	142 Soldiers	? Horses

Preston Barracks Accommodation from 1863 War Office Plans[10]

1863	Constructed for	18 Officers	404 Soldiers	230 Horses

Preston Barracks in the 1841-1891 Census Returns[11]

1841	13 Officers	243 Soldiers	(Royal Scots Greys)
1851	17 Officers	265 Soldiers	(Princess Royal's Dragoons)
1861	13 Officers	340 Soldiers	(5th Dragoon Guards)
1871	? Officers	? Soldiers	(6th Inniskillings)
1881	1 Officer	59 Soldiers	(5th Dragoon Guards)
1891	11 Officers	351 Soldiers	

The government's report on Preston's accommodation in 1848 provided more information on the development of the Cavalry Barracks and the facilities there. Whilst it made no further mention of weather-boarded timber huts similar to those illustrated at Church Street (perhaps the last remnants of these original buildings were lost in a substantial fire that swept through part of the barracks in 1837) by 1848 the original 33 huts had developed into a row of three, multi-roomed, brick built blocks each coated with white lime inside and out and covered by a slate roof. The block to the north had twenty two individual barrack rooms, each 17ft wide by 27ft 9 inches long and just over 9ft in height. Every room had two windows and a fireplace and although regulated to hold 8 men, usually accommodated

11, including some soldiers' wives and their children. A further number of troopers also slept in the vaulted roof space over the stabling. As common practice in other barracks, matrimonial privacy was achieved by simply draping an old blanket over a cord. Being accepted as part of the barrack room community, soldiers' wives were subject to military discipline and often became unpaid orderlies whose duties included tidying up after all the others and cooking. Twenty boilers were provided in the single kitchen, which everyone was expected to take turns to use. No provision was made for the enlisted men to wash, but one room in the barracks was fitted with two large coppers so uniforms at least could be laundered.[12] The officers' accommodation was much more comfortable, better furnished and far less crowded. Thirty officers shared ten furnished rooms, which although 1ft 9 inches less in width, were otherwise identical in size to the rooms provided for the enlisted men. Services, such as washing and cleaning were undertaken by the officers' servants or orderlies, although some soldiers' wives also took on these tasks for a small payment.

Information from Newspapers
(The Sussex Weekly Advertiser/Agricultural Express)

There were occasions were a there were noticeable differences between reports found in official government papers and the local media. The latter had a tendency to over-estimate or dramatise events, sometimes coming up with unrealistic statements, which may have been early attempts at military propaganda. Nevertheless, there always was an underlying element of truth in the news the County papers published.

1795-1796	Preston Barracks was built on substantial brick foundations, but some buildings were still only clad in temporary weather-boarding.
1796 (24th April)	Construction of the barracks' south wing stopped due to a shortage of bricks.
1796 (17th Oct)	The weather-boarded house, originally the canteen, was now to be used as the Barrack hospital.
1800 (February)	It was decided to make the Barracks permanent.
1803 (June)	Barrack accommodation to be extended for 670 Cavalry soldiers.
1810	Racket courts to be provided at the expense of the 10th Dragoons.
1812	Funds of £26,300 to be made available for new stables and enclosing the barracks behind a high wall.
1818	Stabling for 400 horses along the Lewes Road to be demolished and a high brick wall erected in its place.
1859-1867	Further unspecified additions to the barracks.
1870-1871	£1,500 allocated for the reconstruction of soldiers' barrack rooms.
1874-1875	£5,000 allocated for the new riding school and infirmary stables.

n.b. Whilst newspaper reports like those above certainly provide a guide to the alterations and additions proposed for the barracks, it should be remembered that not all proposals for barrack refurbishment were undertaken

Other documents in the Public Record Office also referred to further construction projects which took place at Preston Barracks between 1853-1858 (The Crimean Blocks) 1875-76, 1882-83, 1893-95 & 1897-98. Building projects were also mentioned in a short résumé of the history of Preston Barracks published in the Brighton & Hove Herald on 22nd September 1956. The Herald reported that major alterations were undertaken between 1859-1867, a whole block was rebuilt in 1887 and various other parts of the barracks were upgraded between 1899-1900.[13] In all probability, the works referred to on the final date were the improvements within the barracks during the Boer War, including the provision of a new three bedroomed house for a Warrant Officer. The latter, shown below, was erected next to the Crimean blocks in the southern part of the compound, alongside the stables, guard house and the magazine.

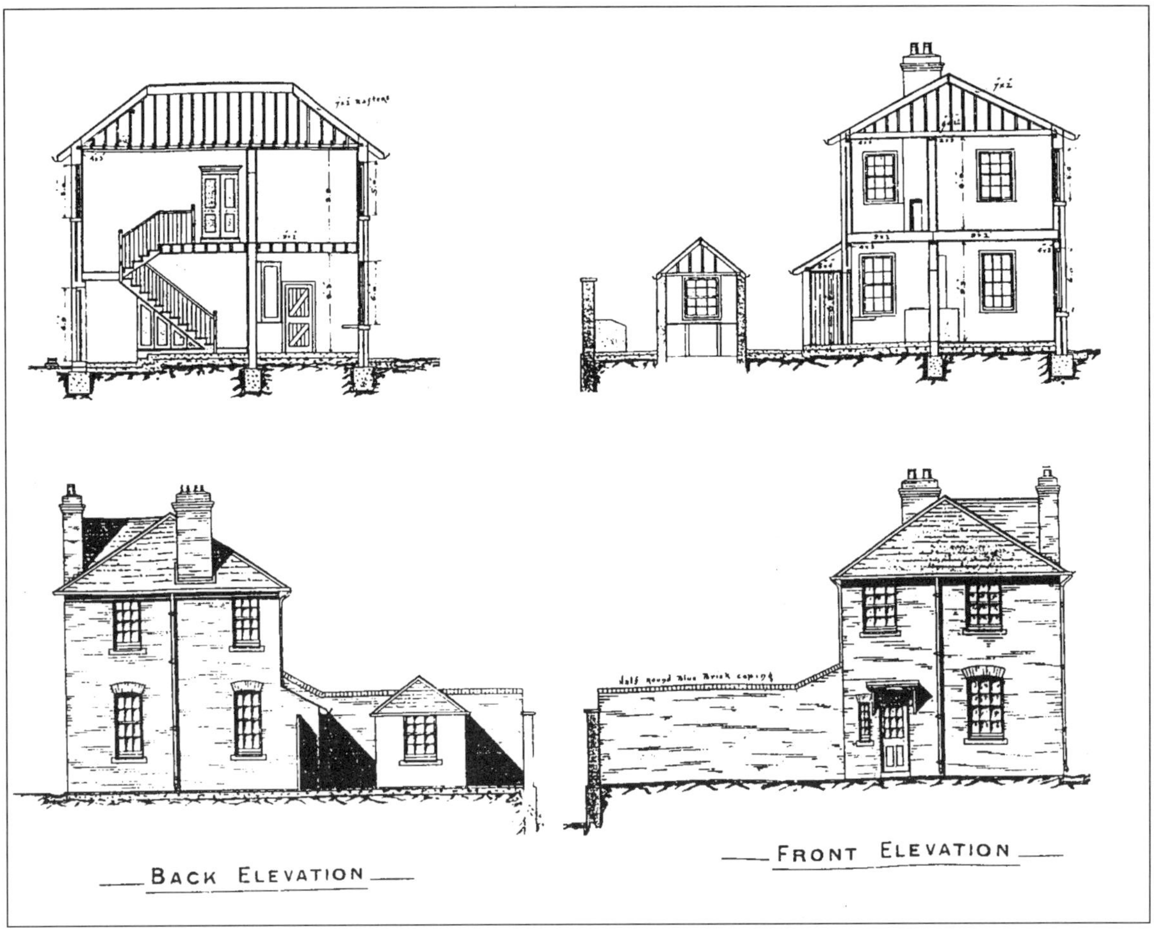

The New Warrant Officer's Quarters (1899)

This increased activity at the end of the 19th century was a reflection of how important Brighton had become as a staging post for soldiers being sent to participate in numerous conflicts throughout Britain's extensive Empire and in the many other countries where the state had a commercial interest.

Barrack Water Supplies

As yet no documents have come to light that detail how Preston Barracks obtained their first supplies of water. It has to be assumed that in 1795, the entire barracks was serviced by a single well, for an entry in the records held by the local water company refers to the antler of a long extinct reindeer being found in the chalk when the first barrack well was dug. By 1835, the number of wells within the barrack compound had increased to six, all over 60ft. deep. Five of these were equidistantly spaced between the main barrack blocks and the western wall. The sixth was outside the barrack hospital on the northern perimeter. Even with six wells in the barracks, water supplies were unreliable, for Brighton's high demands frequently caused the barrack wells to dry up as the water table dropped. The main problem was that much of the water that passed in the subterranean stream beneath the barracks was being drawn off by the powerful steam engines of the Hollingdean based Brighton, Hove & Preston Water Company. The water company which went into production in 1834, was directly next to the barracks and had sunk a 12ft wide well to a depth of 97ft. It used two 8 inch bore steam driven pumps to extract almost all of the water and pumped supplies to its own subterranean tanks and a large reservoir high on the hills above the town. In 1835, the resulting water shortage in the barracks caused military officials to consider following the water company's example and they proposed that military engineers would sink their own shaft to a depth of 228 ft at a position some 140ft up Hollingdean Hill. The shaft was to be capped with an ornamental tower containing a small steam engine which could then pump supplies into a holding tank. From there, the water would be fed by gravity to strategic sites within the barrack complex. This ambitious plan was never put into effect, and was replaced by the less expensive and more unreliable alternative of lowering the central well outside the forage barn which cost the more moderate sum of £30.19s 11¾d.

In 1838, there were further water shortages requiring each of the five main wells in the barracks to be sunk an additional 10ft for a total cost of £124.6s.0d. Whilst the work was in progress, the Barrack Master was instructed to obtain water on a short term basis from the Hollingdean Pumping Station which was only 261 yards away. By 1847, the primitive winching of water from the wells for men and horses had ceased and more conventional hand pumps were installed over several well shafts.[14] A plan of the barracks from 1863 shows an engine room abutting the officers' quarters, which implies that the idea originally put forward in 1835 requiring the installation of a steam pump was finally looked upon more favourably.

Worship

Building a barracks in Preston created two separate communities within the parish. Whilst, most of Preston's 200 rural villagers lived within a short distance of the parish church which was on the westerly route to London, a separate and far larger military community was being built up 1½ miles due east, around the barracks. Whilst some soldiers still preferred to make their way to the old Brighton parish church of St. Nicholas, up until about 1840 the villagers and the majority of barrack community all came together to worship in Preston's parish church of St. Peter. The many military entries in St. Peter's parish registers provide a demographic pattern which gives an insight into the realities of barrack life, showing it to be hard, frequently short and with a high rate of infant mortality. The Preston cemetery still has several monuments to brave soldiers from the barracks who survived the Battle of Waterloo, the Charge of The Light Brigade, or in the case of one unfortunate individual, died after falling from his horse during a local exercise.[15]

Inevitably, the tremendous influx of people and military hardware into Preston's formally small and quiet rural community resulted in rapid suburbanisation and social change as the eastern precincts of the parish rapidly became engrossed into the expanding urban sprawl of Brighton. To cater for the increasing population of the two communities, a newer and far larger church also called St. Peter's, was built on the Brighton section of the Lewes Road. Many of the soldiers found this church more accessible and by 1841 much of the military contact with the smaller family church of St. Peter's in Preston began to dwindle. Over the years further changes took place. In the 1870s for instance, the 4th Irish Dragoons discovered that there was no convenient Roman Catholic church and began fund raising for a small chapel in Milton Road. Its popularity resulted in the chapel becoming a catholic school and a far larger church of St. Joseph being built at the bottom of Elm Grove.

Events at Preston Cavalry Barracks

Almost certainly some of the early press reports about the barracks are heavily romanticised. The earliest story was published on 23rd August 1796, and told how Preston Barracks acquired its first cat. Apparently an itinerant family had camped in the grandstand on Brighton racecourse and accidentally set fire to it. The flames were so high they lit up the sky for up to 30 miles around and many people thought it was a beacon to inform them that Napoleon's forces had landed and rushed to the scene with any weapon they could grab hold of. Included in their number was a young officer from the 10th Dragoons who had been sent up the hill from the barracks to find out what the military should do. On realising it was a false alarm, he settled down to enjoy the fire and to watch others panic. Suddenly a severely singed cat shot out from beneath the inferno, bounded up onto his horse and affixed itself with its claws almost permanently to his dress uniform. His initial reaction was to shake it off, but after taking pity on the poor animal, he returned with it to his quarters in the barracks, where the Regiment gave it a new home.[16] The Times of 1803 published another story which carried a slight racial slant. It told of the black bugle boy who was confronted by the Prince of Wales as he arrived at the dead of night to inspect his regiment. The newspaper suggested that the boy was so over-awed by his Highness' magnificent presence, that he could not correctly form his lips in order to blow the call to turn out the guard of honour. However, on hearing the commotion made by the Royal party, a far more experienced white bugler arose from his bed and stood to attention at an upper window, blowing a Royal Tribute whilst still in his night attire.

Other newspaper reports were much more factual. In 1810 for instance, the officers from the 10th Dragoons did provide all the funding to build their own racket courts, and these survived for almost a century. They can clearly be seen in the 1863 barrack plans, but by 1912 the courts had been relegated from recreation and were being used as coal bunkers for the barrack furnaces. The most informative reports and first hand accounts however, appear in the memoirs of those who were present at the time. Individuals such as Alexander Somerville who served in the barracks in 1832 and Frederic Ogburn who recalled his time around the barracks in his schooldays.

In 1832, Somerville enlisted as a trooper in The Royal Scots Greys and was posted from London to Preston Barracks. As was common practice, he and another new recruit called Ireland had been marched under military escort to Preston, pausing overnight in the substantial barracks in Horsham. On reaching Preston village, they were directed to the barracks over two large hills which nowadays are known as The Droveway and Hollingdean dip. Although it was a cold day in January, they were both met at the gates by numerous local people who were offering to buy everything they wore for a fraction of their value. At that stage Somerville had not fully appreciated that with cholera sweeping England, Preston Barracks were effectively in isolation and all those admitted were under strict orders to destroy everything that had been in contact with the community outside. He should have realised that something was to be gained when the sentry surreptitiously asked if he had anything he valued brought in, but being new and unaware that some military personnel have a hidden agenda, he thought little of it. On receiving no response, the sentry returned to the louder military tone, and instructed him to report to the barrack hospital where he would be staying. As it was commonly believed that contact with a cholera carrier would almost certainly result in being infected, there was to be no communication with other soldiers in the barracks. Therefore in the short journey across the barrack compound to the hospital in the northern most corner, few soldiers spoke, save for the handful who called out from the safe distance of the high windows. Once there, Somerville and Ireland were escorted into an isolation ward where they found James Miller, who almost certainly had cholera and was in a state of delirium and agony. Although only recruits, they were ordered to strip so that their infected clothes and other possessions could be burnt, given light blue woolen robes to wear, and then instructed to care for Miller's needs. Somerville watched as his almost new clothing was removed from the room on the end of a pitchfork and taken into the yard in the rear where it was to be destroyed. On looking out of the window, he was astonished to see that in reality, straw was being burnt instead of his clothing and then he realised that the garments were actually being sold to the same black-marketeers that he had avoided on the barrack perimeter. Noticing that he had been observed, the orderly quickly returned with a small bottle of brandy and in a broad Scottish accent advised Somerville to keep his mouth shut. Later, Somerville became very aware that barrack rules were constantly being bent and throughout the cholera epidemic, things were constantly being smuggled into the compound. The biggest item he heard of was a set of bagpipes which arrived concealed beneath the hoops of a skirt worn by a serving soldier's wife.

When Miller recovered, he returned to being the servant of an eighteen year old officer but made sure the two troopers who had assisted him through his ordeal had opportunities to acquire extra income and passed on some of his tasks so they could earn an additional sixpence. As they were also now certified as free from possible infection, Somerville and Ireland had their hair trimmed to the Regimental length (still very long by today's military styles) and were supplied with the following items of uniform and kit which a cavalry trooper in Preston Barracks needed to carry out his duties.

A Cavalry Trooper's Kit

Overalls	Stable Jackets	Regimental Dress Coats
Boots	Forage Cap	Leather Stock
4 Linen Shirts	2 Flannel Waistcoats	2 Pairs of Flannel Drawers
4 Pairs of Worsted Socks	2 Pairs of Gloves	1 Pair of Gauntlets
A Curry Comb and Brushes	A Horses Main Comb Sponges	
Soap	Bath Brick	2 Towels
Button Stick	Button Brush	Rot Stone to clean buttons
Boot Brushes	Blacking	Clothes Brush
Brush Bag	Horse Cover Cloth	Horse's Nose Bag
Corn Sack	Saddle Bags	Military Cloak
Turn Screw	A Picker (for horse's hooves)	2 Pairs of straps for the overalls
An account book with printed regulations		
A "Save All" containing:	Knife, Fork, Spoon, Razor, Comb, Shaving Tackle	

Having been shown in meticulous detail how to put the uniform on and use all the equipment, Somerville was allotted to one of six troops, lettered from A—F. He found himself in "D" Troop which was commanded by Sergeant Major Gardener, who was one of the very few soldiers left with the regiment who actually fought at Waterloo.

Somerville soon found out that nothing was free and that every item of the kit he was issued with had to be paid for. All but 10/- of the £2.12s.6d. bounty which had induced him to sign on was deducted towards this cost, which left him owing the regiment £3.10s.0d which had to be paid off over several years. He did receive 1s.4d. a day as a trooper in the Cavalry (unlike the Infantry who only earned 1s.1d a day) but that was before many stoppages were made from it. Rations cost up to 8d a day and vegetables cost a penny extra. In addition, every soldier was duty bound to take pride in himself and more importantly, keep his horse in immaculate condition. This regulation meant that he often had to supplement the above issue of kit out of his own pocket and continually put money aside for those unforeseen repairs to essential equipment. The net result of all these numerous deductions and constraints was that his actual disposable income dropped to around 2d a day, and just like all the other troopers in his barrack room, he was permanently in debt.[17]

Frederic Ogburn also recalled Preston Cavalry Barracks around 1840 when he was only a child. His major reminiscence was of an officer based at Preston who went to Brighton Races and embarked on a gambling spree at the gaming tables there. After losing heavily, he felt he had been cheated so he returned to the barracks, armed his troop with sticks and other weapons and marched the men back to do battle with the gypsies who ran the tables. Ogburn wrote that on that day many of the spectators turned their attention away from the races to the enjoy watching the fracas, until a rather large number of police reinforcements arrived to stop it.[18]

Prince Albert's Own Hussars

In John Wright's research into the history of 11th Prince Albert's Regiment of Hussars, he discovered that as well as being at the Battle of Waterloo in 1815, between 1840-1841, the Regiment was also billeted in Preston Cavalry Barracks, Church Street and at the Royal Pavilion Stables. When they first arrived on 26th June 1840, many soldiers thought the town to be dull, but in their brief 10 month stay, they added considerable colour. The aspect which endeared them most to the local community was the regimental band which played regularly all over town, appearing on the Chain Pier and the bandstand in Steine Gardens, at numerous private functions and even heralding the opening of Brighton's new railway station. Frequently they assembled outside number 45 Brunswick Square, where their commanding officer, The Earl of Cardigan had his private residence.

On 3rd September 1840 whilst in Brighton the soldiers were issued with brand new uniforms. So startling was its effect when the regiment first paraded through Brighton, they were immediately given the new and affectionate nickname of, "The Cherry Pickers." John Wright wrote that the uniform consisted of,

> *...a colbec or busby of seal skin (undress caps were of crimson cloth, coronet shaped, with a gold tuft and gold band), a jacket of Prussian Blue with elaborate gold braiding, blue pelisse with crimson lining, cuffs and collars of sable fur and bright [cherry] red trousers with broad yellow seam stripes. The horses' labraque was crimson bordered with gold lace, decorated with imperial crowns, the initials of the Queen and Prince Albert, the regimental emblem, and the names of famous victories in which the regiment had figured.*[19]

Whilst the regiment was popular within the local community, their Commanding Officer, the Earl of Cardigan was not, he was even disliked by many of his own officers. Their general opinion was that the Earl, who had a tendency to be dogmatic and very determined to get his own way at whatever cost, had reached his exalted position within the regiment by the power of his purse rather than the wealth of his experience and several officers with far greater military skill took exception to this. After a while, one of them, a Captain Reynolds, felt he had endured the Earl's insults for far too long and passed him a note saying that had he not been his Commanding Officer, he would have challenged him to a duel. Using his authority to the full, the Earl immediately countered by convening a Court Martial in Preston Barracks' hospital and had the Captain cashiered. Many knew that there was much more to it than just a captain being insubordinate, but nevertheless, the unfortunate man was drummed out of the regiment. An interesting supplementary note is that throughout their stay in Brighton, the regimental cavalry from the 11th Hussars regularly practiced mock charges on the local racecourse and the skills they learnt were later put to good effect in battle at, "The Charge of the Light Brigade." Although nothing should be taken away from the considerable heroism shown by the men, the losses experienced must reflect certain aspects of the inadequacy of their leadership, and by being drummed out of the Regiment, the former Captain Reynolds may have inadvertently owed his life to the Earl!

Going on Leave From Preston Barracks

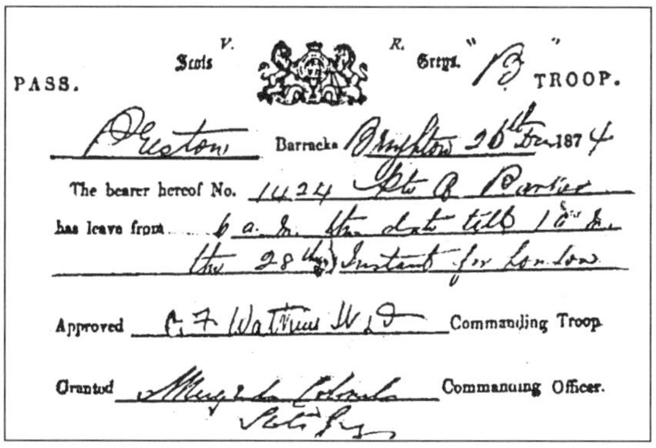
PASS. V. R. Scots Greys. "B" TROOP.

Preston Barracks Brighton 26th Dec 1874

The bearer hereof No. 1424 Pte A Barker

has leave from 6 a.m. the date till 10 a.m. the 28th Instant for London

Approved [illegible] Commanding Troop.

Granted [illegible] Commanding Officer.

Leave Pass from Preston Cavalry Barracks

Just like today, every soldier who left any barracks needed to have official sanction. The pass opposite was issued to Private 1424, A Barker,from B Troop of the Royal Scots Greys on 26th December 1874, presumably to enable him to get to London in order to participate in the seasonal festivities. However, occasionally soldiers did abscond and the local police were fully aware that all soldiers off limits should possess a pass and that if the man could not produce it, they could arrest him.

In researching his own background, fellow family historian, John Charmers discovered the following article published in December 1886 detailing how a policeman ancestor of his detained a soldier from Preston who was on the run.

> *At the Justices' office on Monday, before Henry Upton Esq., John Louder belonging to R Troop of the 5th Lancers stationed at Preston Barracks Brighton, was brought up in custody on suspicion of being a deserter from the above regiment. P. C. Charman of Tillington met Louder in uniform near the town on Saturday afternoon last and asked him for his pass. He said it was in his greatcoat , so a constable was sent for it , but one could not be found. The prisoner then admitted being absent without leave and was committed to Preston Barracks to be handed over to his Commanding Officer. It has since transpired that he had no pass and had deserted from his regiment on 13th November last.*[20]

It is expected that Louder was then taken under armed escort to the cells in Preston Barracks.

In more recent times, the same but largely unchanged cells were etched into the memory of the unique English actor, poet and comedian, Spike Milligan. Spike's light-hearted reminiscences do not dwell on the grandeur of the building, but reflect more humorously about the isolation of his own incarceration in Preston Barracks' Victorian guardroom where he had the misfortune to be detained for two weeks during the second world war. During that time he had ample opportunity to make a detailed study of his relatively small, enclosed environment and recalled that each cell was approximately 20ft high, had a stone floor measuring about 6ft x 7ft and was only furnished with an iron framed bed. There was no light except that which came from a small window way up near the ceiling and to impede his escape it had a single iron bar down it. Each door was made of 2 inch thick solid iron and was fitted with a peep hole which was covered by a metal flap swinging from a pivot on the outside. He knew the cell door well, for his punishment included oiling the lock and hinges, then sand-papering, priming and undercoating it, before finally repainting everything gunmetal black. Quite often he was allowed outside into the Victorian guardroom which was also very basic. It had a small verandah beyond which was an exercise yard that occasionally had to be weeded by hand. A special privilege was to be allowed out on the railway embankment to pick blackberries for the sergeants' mess.[21]

Photography

With photography continually growing in popularity throughout the Victorian era, soldiers were a good source of income, many of them using the services of Mr. Charles Grey who in 1885 set up as a Military Photographer in the Lewes Road near Preston Barracks. Although soldiers often made their way to his studio, few photographs ever seem to have been taken within the barracks. Two which have been discovered are both of the 4th Royal Irish Dragoons. The earliest, from 1885, was submitted to the Brighton and Hove Gazette in 1975 by Mr. Peter Turrell, then of 80 Whitehawk Avenue. It showed the Dragoons gathering, before being dispatched to Liverpool docks where they were to embark on a troopship to fight in the Egyptian campaign. Peter reported that his father (also called Peter) was eighth from the left in the middle row.

The 4th Royal Irish Dragoons (1885)

The uniforms with ceremonial swords and typical pill box hats of 1885 make an interesting contrast with another photograph from Tony McKendrick Warden's collection which shows men from the same Regiment wearing the more formal uniform which had come into use by the turn of the century. Their daily ceremony to change the guard always drew enthusiastic onlookers, for small boys, pretty girls, proud parents and old soldiers frequently gathered at the barracks' pedestrian gate to watch. A curiosity is the double sided post box inset into the barrack wall, for it would seem to be keeping up the postal service that was in operation almost half a century earlier. The illustration of the toll gate keepers cottage which had formerly been just on the other side of the wall, indicates that that too had a post box which hung by the door.

The Changing of the Guard by the Royal Irish Dragoons (1904)

The high wall and wide gates have now gone, and bellowed commands and clattering boots can no longer be heard. The only parade still to be seen in the area is that of the row of very recognisable shops and houses in Coombe Terrace which remain in situ opposite the former site of the impressive Cavalry Barracks that were once in Preston.

References

1 *Sussex Weekly Advertiser* (WSRO)
12th October 1795 — All the references in (1) are referred to in A Hudson's *Napoleonic Barracks in West Sussex* MP 2549
28th March. 1796
25th April 1796
9th May 1796
4th July 1796
17th October 1796

2 S & J Farrant.
Preston in the 17th & 18th Centuries 1975 page 16

3 Hudson A *Napoleonic Barracks in West Sussex* MP 2549 WSRO
Refers to papers in Kew Public Record Office WO 40 13

4 Attree H *Topography of Brighton* 1809 page 43

5 The Brighton Ambulator 1818
Wallis's New Brighton Guide 1826 pages 40-41
Brighton As It Is 1831 page 66 1832 page 54 1838 page 56
A Stranger's Guide to Brighton 1839 page 35
Taylor's Royal Guide to Brighton 1832 page 152

6 1858 Plan of Hollingdean Waterworks ESRO S.W.A. 4/4/6

7 Early War Office Papers from the Public Record Office in Kew
July 1794 WO 40 07
February 1800 WO 40 13
1802 WO 30 81 page 44
June 1803 WO 30 37 page 145
1804 WO 01 88 page 44
6th March 1806 WO 40 25 page 9

8 Ann Hudson refers to *Sussex Weekly Advertiser*
4th September 1801
21st September 1801

9 *House of Commons Return of Barracks*
25th March 1812
5th January 1822
2nd March 1832
31st March 1847

10 James Col. H *Plans of the Barracks etc., in the Brighton Division of the South Eastern District* 1863

11 Census 1841 HO 1112 E.D. 10+ Folio 16
Census 1851 HO 107 1647 Folio 142
Census 1861 RG 9 604 E.D. 3
Census 1871 *Barracks not found on the Preston Census.*
Census Church of the Latter Day Saints 1881 1103 101 2nd page 7374

12 *House of Commons Return of Barracks*
2nd March 1832
31st March 1847

13 *Brighton & Hove Herald*
22nd September 1956

14 *House of Commons Return of Barracks*
2nd March 1832
31st March 1847

15 Grant R C Military Baptisms, Marriages & Burials at the Church of St. Peter, Preston Village, Brighton, Sussex. (1793—1840) 1997

16 Erridge J *The History of Brighthelmstone* 1862 page 282

17 Somerville A *Autobiography of a Working Man* 1848 pages 197-205

18 Ogburn F *Eighty Years of Reminiscences* 1914 page 18

19 Wright J *Sussex County Magazine*
Vol. 28 Brighton to Balaclava pages 473-475

20 *West Sussex Gazette* 2nd December 1886

21 Milligan S *Gunner Milligan,* 954024 1995 pages 36—39

22 War Office Papers from the Public Record Office in Kew
20th May 1835 W0 55 792

PRESTON CAVALRY BARRACKS' HOSPITAL

Initially I assumed that the demolition of Preston Barracks had totally eradicated every building of historical interest, but that thought subsequently proved to be incorrect. A study of plans from 1838 and 1898 revealed that at least one original building from 1793 still remained within the barracks. Although altered internally by frequent efforts to make it more suited to changes of use, the former barrack canteen which had been re-designated as a hospital in 1800, still stood on the same foundations in the far north west corner of the barrack compound. It was there that Alexander Somerville was incarcerated for several weeks whilst he saw off the 1832 cholera infection and where in 1840 Captain Reynolds had his court martial. Now used by the cadets, there is no evidence of its weather-boarding or whitewashed frontage, but the remains of the front steps which provided access from the east and the two very distinctive recesses which dominated the western facade are still clearly visible.

Sketch Plans of the Former Barrack Hospital.

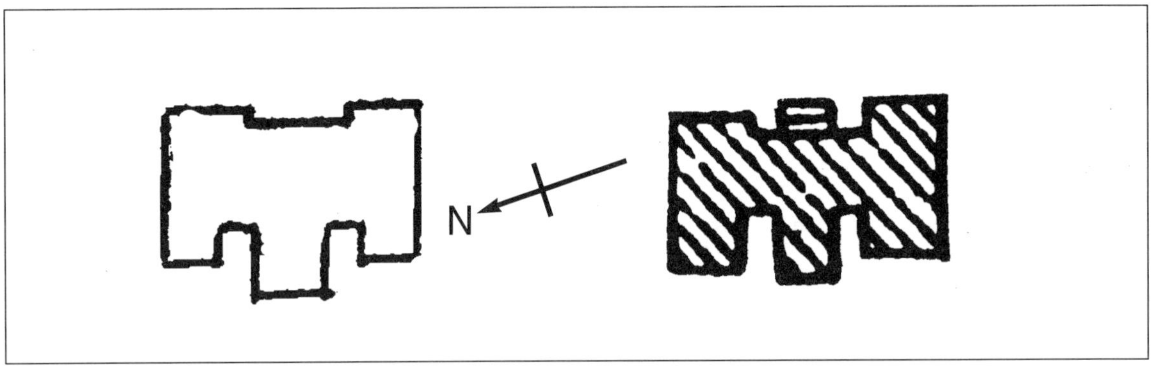

Left: from 1838 plan of Barrack wells. ***Right: from 1898 Ordnance Survey***

The hospital well shown on the 1838 plan is now infilled, but in all probability it still exists just beneath the surface of what is now a grassed area to the north of the building.

Present Day Photograph of the Former Barrack Hospital

There are no plans available to illustrate how the hospital was laid out, but a general schedule from May 1835 suggests that there were three main areas, a kitchen, a surgery and a number of wards. In true military fashion, the regulation was that each of these rooms had to be furnished with the following equipment which hospital orderlies were expected to account for and continually maintain in good order.[22]

The Kitchen:

One Grate or Range	Four Meat Dishes	One Ladle
One Shovel	One Porringer (per patient)	One Flesh Fork
One Poker	One Plate (per patient)	One Trivet
One Pair of Tongs	One Candlestick	One Boiler & Lid
One Fender	Two Washing Tubs	Two Mops
One Table	Six Saucepans	One Hair Broom
Two Forms	Two Tea Kettles	One Round Towel
One Hand Scrubbing Brush	One Quart Pot	One Coal Box
One Heath or Birch Broom	One Pint Pot	One Coal Tray
(changed weekly)	Six Water Pails	One Tin Lantern

The Surgery:

One Candlestick	One Grate	One Table
One Round Towel	One Shovel	Two Chairs
One Slipper Bath	One Poker	One Coal Scuttle
One Shower Bath	One Pair of Tongs	One Fender

In Each Ward:

One Table	One Coal Tray	One Candlestick
One Form	One Grate	One Bed Pan
One Close Stool & Pan	One Shovel	One Urinal
One Hair Broom	One Poker	Three Spitting Cups
One Dry Rubber or Long Scrubbing Brush		One Pair of Tongs
The Correct Number of Window Blinds		One Fender

For Each Patient:

One Bedstead	One Sacking Bottom	One Gown
One Bolster	One Sacking Cord	One Waistcote
One Palliasse	One Bolster	One Trousers
Two Blankets	One Rug	One Cap
Two Sheets	One Chamber Pot	One Pair of Slippers

Epitaph

It is most fortunate that Preston Cavalry Barracks still has its original hospital. Built as a canteen in 1793, the walls initially would have echoed to boisterous mealtime conversations and later, to exciting recollections of injured soldiers returning from the many battles undertaken to form and protect the British Empire. Oblivious to its past, in more recent times, I unknowingly used this historical building myself when I was a soldier in the Territorial Army, and now I hear the walls vibrate to the curious youthful banter of army cadets. Hopefully I have now put it on record that the former hospital in Preston Cavalry Barracks is the only remaining 18th century military building within the Brighton area to survive the speculator and redevelopment. Clearly it is part of our local military heritage and we should ensure that it remains with us for some considerable time to come.

A FINAL THOUGHT FROM THE AUTHOR

By the beginning of the 20th century, all the evidence of Brighton's military past had been removed from the centre of town and it was only Preston Cavalry Barracks which desperately clung to the last remnants of former Regency and Crimean architecture. With new wars to fight and new regiments to accommodate, Preston continued to preserve many aspects of 19th century romance whilst still having to cope with the harsher realities of 20th Century warfare. Although much of the exterior architecture was kept an ever increasing number of Preston's former livery stables were converted to garages, forcing the Regiment's horses to share with motorised transport. As the mounted Cavalryman or Lancer moved into the new world of tank warfare, the nature of the Barracks and the troops lodged there changed. Newer brick built buildings sprung up on the former parade ground, sandwiching the old historic uniqueness of the Regency barracks in the void between them and the high railway embankment. Although hidden from public view, most of these original buildings were still preserved, until almost the end of the 20th century, by which time they had witnessed many advances in military equipment and even survived fire bomb attacks by German aeroplanes. It is therefore most regrettable that just a few years before the Preston Barracks' second centenary, pressure for a local commercial site in the immediate area resulted in the barracks being sold off for development. By 1990, much of the historic site had been cleared, and the small boys who had once gazed through the iron railings at uniformed soldiers and military vehicles now contented themselves by watching hard hatted demolition workers unceremoniously going about their legitimate business. A fact that returns me with some considerable sadness to my start point.

When I commenced collecting material to help unravel some of the mysteries of Brighton's military heritage, I was fully aware of the vastness of the subject and my own limitations. To conduct the painstaking research needed to assemble such a comprehensive dossier would need an inordinate amount of funding and I knew that my time and available resources were inadequate. Furthermore, knowing that the illustrations and photographs which I had access to were mostly over 100 years old, I wondered to what extent public interest might be reduced because of their poor condition. In essence I sought to present something which included sufficient pieces of an extensive jigsaw to better define the outer framework of Brighton's military past and by doing so, capture the interest of many readers with further archive material. With luck, this document might just motivate readers to embark on their own collections of military data, which in turn would further augment the study of local history, and help to fill in those numerous areas where there are frequent gaps.

As one lasting reminder of Preston Barracks, I have included the better examples of the photographs that first started me on this quest, showing the last moments before the forlorn shells of a unique collection of distinctive 19th century military buildings were removed. Local people have lost the last remnants of their military heritage, not to a foreign invader, but to what has been suggested by some as being a more economic use of the environment. As occasionally happens, the pursuit of economic viability has left local people helpless whilst officialdom sanctions the clearance of important aspects of their past. It is my sincere hope that in the future, all administrators will become more aware of the wider consequences of their actions and take every opportunity to ensure that all important architecture is at least fully documented or better still, preserved. There must be numerous other Brightonians like myself, who place considerable value on their heritage, much of which lies encapsulated in the many unique buildings of historical interest which still survive in the more remote corners of the town.

The Crimean Block During its Temporary Reprieve
Photograph reproduced by the kind permission of Edward Start of Open Spaces

The Final Demolition of the Original Barrack Buildings
Photograph reproduced by the kind permission of Edward Start of Open Spaces

BIBLIOGRAPHY AND SOURCES OF ILLUSTRATIONS

Bibliography & Other Associated Reading

Attree H. R	*The Topography of Brighton*		1809
Attwick W H	*The Jubilee of Brighton Corporation*		1904
Jane Austen	*Pride and Prejudice*		
Bishop J G	*The Brighton Pavilion and its Royal Associations*		1903
Barnes Major R	*A History of the Regiments and Uniforms of the British Army*		1972
	The Brighton Ambulator		1818
	Brighton and its Environs		1830
	Brighton As It Is		1831 1832 1838
	The Brighton Gazette		
		16th April	1832
		12th June	1862
	Jubilee Supplement	30th June	1904
	The Brighton & Hove Herald	22nd Sep.	1956

The Brighton Reference Library & Museum (Brighton)

	The Brighton Rate Book *An Equalisation of the Parish*		
	Houses, Stables, Coach Houses and Other Buildings		1814
	Etching of the Execution of Cooke & Parish		1795
	The Library's Collection of Local Photographs		
	Montgomery J *Murder in Church Street Barracks Undated*		
	Painting by C Wing or De La Motte of Preston Barracks		c. 1850
	Views and Reviews	August	1896
		September	1896
		October	1896
	Woodard H. J. M. A. Thesis *The Volunteer Reviews of 1861-1879*		1992
Cannons Services	*Historical Records of the 10th Prince of Wales'*		
	Own Royal Regiment of Hussars from 1715-1843		
Census	Church Street Barracks Census Returns		1841-1861
	Preston Cavalry Barracks Census		1841-1891
Clark S	*English Folk Dancing*		1956
Dale A	*Brighton Town, Brighton People*		1976
Edwards D	*Soldiers in Revolt An undated political pamphlet*		
Erridge J A	*The History of Brighthelmstone*		1862

East Sussex Record Office (Lewes)

	ADD MSS 5440 No 259 Letters to Lord Pelham		1798
	Plan of Hollingdean Waterworks S.W.A. 4/4/6		1858
	Shifner MS *3009 Regulations for Militia*		1852
Farrant S & J	*Preston in the 17th and 18th Centuries*		1975
Goodwin J E	*The Military Defence of West Sussex*		1985
Grant R C	*Military Baptisms, Marriages & Burials at the Church of St. Peter,*		
	Preston Village, Brighton, Sussex (1793—1840)		1997
Hannington J	*The Personal Journal of "Bishop" James Hannington*		
Hudson A	*Napoleonic Barracks in West Sussex*		
	Illustrated London News	Easter	1859-1872
Kitchen F	*Sussex History* No 21 Brighton's Batteries	May	1986
Kitzmiller J.M	*In Search of Forlorn Hope*		1988
Laws M. Lt. Col.	*The Royal Artillery Journal* Vol. 84	October	1957
	The Graphic	11th April	1885 page 361
Medhurst & Pitcher	*Rolls of Honour & War Memorials in*		
	Some Villages & Towns in East & West Sussex. Vol. 8		1994

Melville L	*Brighton History, Follies, Fashions*		1909
Milligan S	*Gunner Milligan, 954024*		1995
Musgrave C	*Life in Brighton*		1981
Ogburn F	*Eighty Years of Reminiscences*		1914
Public Record Office (Kew)			
	Correspondence with Southern Command		
	WO 40 07	July	1794
	WO 40 13	February	1800
	WO 30 81		1802
	WO 30 37	June	1803
	WO 01 88		1804
	WO 40 25	6th March	1806
	WO 55 792	20th May	1835
	WO 13 700-725	20th May	1835
	House of Commons Return of Barracks		
		25th March	1812
		5th January	1822
		2nd March	1832
		31st March	1847
	Col. H James' Plans of the Barracks etc in the Brighton Division of the South Eastern District		1863
Roberts W	*Francis Wheatley R.A, His Life and Works*		1910
Saunders W	*A Strangers' Guide to Brighton*		1839 & 1852
Scott H	*The English Song Book*		1925
Somerville A	*Autobiography of a Working Man*		1848
Sussex Archaeological Collections			
Volume 39	*The Rediscovery of the Old Town Well*		1894
Volume 107	Margary I *Militia Camps in Sussex*		1969
Volume 124	Hudson A *Gazetteer of Barracks in Sussex* (1793-1815)		1986
	Sussex County Magazine Vol. 28 Wright J Brighton to Balaclava		
Sussex Weekly Advertiser		18th Nov.	1793
		30th Dec.	1793
		12th Oct.	1795
		28th March	1796
		25th April	1796
		9th May	1796
		4th July	1796
		17th Oct.	1796
		14th Sept.	1801
		21st Sept.	1801
	Taylor's Royal Guide to Brighton		1832
	The Times	October	1793
Tuppen A. E.	A. E. Tuppen's personal collection of papers from:		
	The Easter Monday Review & The First Sussex Volunteer Meetings of 1865		
	Wallis's New Brighton Guide		1826
	West Sussex Gazette	2nd Dec.	1886
West Sussex Record Office (Chichester)			
	Records of the Royal Sussex Rifles RSR MS 4/2		1862

Illustrations and their Sources

Cover

The Volunteer Review at Brighton.	*The Graphic*	11 April 1885	Page 361

Introduction

An Attack on Brighton by the French Fleet (1545)	Brighton Reference Library		
The Remains of the Brighton Blockhouse in 1761	Brighton Reference Library		
Plan of the Great East Street Battery: (1759-1786)	Goodwin J E *The Military Defence of West Sussex*	1985	Page 56

Early Encampments (1793-1803)

The "Downs" Camp.	Goodwin J E *The Military Defence of West Sussex*	1985	
"Brighton in a Bustle"	Postcard from Brighton Museum		
The Execution of Cooke and Parish.	Etching held by Brighton Museum		
The Encampment at Brighton	Roberts W. *Francis Wheatley R.A. His Life and Works*	1910	
Departure From Brighton.	Roberts W. *Francis Wheatley R.A. His Life and Works*	1910	

The Music to Brighton Camp Clark S English Folk Dancing 1956

Later Encampments (1860-1872)

(All illustrations from "The Illustrated London News" unless otherwise stated)

Troops Arriving at Brighton Station	23 April 1870	page 429
March from Brighton to the Assembly Point	30 April 1870	page 448
The Muster on the Level	29 April 1865	page 396
Preparing for the Field	30 April 1870	page 440
Man-handling the Field Guns up onto the Race Hil	13 April 1861	page 355
The Prince of Wales Reviewing theTroops	14 April 1866	page 349
The Mock Defence of Ovingdean	30 April 1870	page 440
The Mock Battle of Rottingdean	30 April 1870	page 445
The Middlesex Regiment Returning to London (Via Patcham)	30 April 1870	page 441

Brighton's Coastal Defences

Plan of the New Eastern Battery: (1793)	Public Record Office in Kew	WO 781/1456	
An artists impression of one Battery in action	Brighton Reference Library		
Plan of the New Western Battery: (1793)	Public Record Office in Kew	WO 781/1456	
The Western Battery	Brighton Reference Library		

Church Street Infantry Barracks and The Royal Pavilion Stables

Roofscape Showing Church Street Barracks (C1841)	Brighton Reference Library		
The Installation of a Stove in the Barrack Store in 1837	Public Record Office in Kew	WO 55/792	
The Infantry Barracks Brighton (1823, 1838 & 1863)	Enlargement of a Silhouette on the 1823 Street Plan by Piggot Smith		
Plans of Church Street Barrack wells,	Public Record Office in Kew	WO 55/792	
Sketch made from Col.H James' 1863 Plan of Barracks			
Former Properties at the Corner of Church St. & Marlborough Place (1834)	Public Record Office in Kew	WO 55/791	page 6
The Two Training Cannons inside Church Street Barracks.	Goodwin J E *The Military Defence of West Sussex*	1985	page 103

Sentries at the North Gate 1832 — Lithograph by T M Baynes after a drawing by J H Good
Plan of the Barrack Accommodation in the Royal Pavilion Stables (1863) — Public Record Office in Kew. Col. H James' 1863 Plans

Drill Halls and Local Volunteers

Gunner W R Wood of the Brighton Corps of the Sussex Volunteer Artillery	Views and Reviews	1896
The Gloucester Road Drill Hall	Own Photograph	
Lieutenant Colonel Charles Smith Hannington,	Photograph kindly loaned by Mr. M. Hannington	
The Officers of the 1st Sussex Artillery Volunteers (1865)	Views and Reviews	1896
Brighton Gun Crew from The 1st Sussex Volunteer Artillery (1865)	Mr A Tuppen's personal collection of papers from 1865	
The Armoured Train	Views and Reviews	1896
Officers of the 1st Sussex Volunteer Rifles in 1862	Views and Reviews	1896
Officers of the 1st Sussex Rifles in 1873	Views and Reviews	1896
Colour Sergeant Edwin Gillam and other Senior NCOs from the1st Volunteer Battalion of the Royal Sussex Regiment	Photograph kindly loaned by Mr. M J Gillam	
Church Street Drill Hall in 1896 and 1996	Own Photograph & Views and Reviews	1896
Colour Sergeant Edwin Gillam	Photograph kindly loaned by Mr. M J Gillam	

Preston Cavalry Barracks

Preston Cavalry Barracks circa 1850.	Original water-colour from Brighton Museum of Art Either De La Motte or C Wing (Speculation over artist)	
1867 Postcard showing the toll gate outside Preston Barracks	Kindly loaned by a member of the Sussex Family History Group	
A simplified sketch of Preston Cavalry Barracks. (Adapted from a War Office plan produced by Colonel H James in 1863 and now property of the Public Record Office in Kew)		
Ordnance Survey of Preston Cavalry Barracks in 1898	Ordnance Survey	1898
Sketch of one of the "Crimean" barrack blocks.	Sketch made from own photograph	
The New Warrant Officer's Quarters: (1899)	Public Record Office in Kew WO 78/3436	
Copy of leave pass from Preston Cavalry Barracks	Source unknown	
The 4th Royal Irish Dragoons (1885)	Brighton & Hove Gazette 12 April 1975	
The Changing of the Guard by the 4th Royal Irish Dragoons (1904)	Photograph from Tony McKendric Warden's collection	

Preston Cavalry Barracks' Hospital

Sketch plans of the former barrack hospital	
Sketches from 1838 plan in PRO Kew WO 55/792 & 1898	Ordnance Survey.
Present day photograph of the former barrack hospital.	Own photographs.
The Crimean block during its temporary reprieve	Photographs kindly loaned by Open Spaces
The final demolition of the original barrack buildings.	Photographs kindly loaned by Open Spaces